Innovation Engines

Innovation in Engineering and Technology Set

coordinated by
Dimitri Uzunidis

Volume 1

Innovation Engines

Entrepreneurs and Enterprises in a Turbulent World

Edited by

Dimitri Uzunidis
Pierre Saulais

iSTE

WILEY

First published 2017 in Great Britain and the United States by ISTE Ltd and John Wiley & Sons, Inc.

ISTE Ltd
27-37 St George's Road
London SW19 4EU
UK

www.iste.co.uk

John Wiley & Sons, Inc.
111 River Street
Hoboken, NJ 07030
USA

www.wiley.com

Library of Congress Control Number: 2017934905

British Library Cataloguing-in-Publication Data
A CIP record for this book is available from the British Library
ISBN 978-1-78630-164-2

Contents

Introduction

Innovation Engines: Entrepreneurs and Enterprises in a Turbulent World

The thesis that the competitiveness of an economy depends on the density and performance of its system of innovation is generally accepted by economists. Nonetheless, it is easy to defend the idea of a correlation between a company's innovation and production capabilities and the wealth of innovation resources in the (national or local) economy in which the business or entrepreneur has developed: namely, the availability of a highly qualified and skilled workforce, large investments in scientific research and technological development, buying markets with significant purchasing power and tendencies to reinvent themselves, suitable institutional frameworks and support, etc. The innovation system is studied like an R&D network made up of scientific and technical institutions, of laboratories for research and engineering with the aim of creating, producing, learning – of innovating. This concept of an innovation system has opened up several avenues for research, two of which have particularly grabbed the attention of political decision makers: the idea of localizing the innovation process, and that of the emergence of innovation networks with technology entrepreneurs acting as the main brokers. The creator of the business that has the knowledge capital is enhanced by the dual investment and (re)appropriation process [BOU 16].

It is not only the entrepreneurial aspects of innovation, but also the economic, financial and sociotechnical aspects that fall within the practices of creating and acquiring the capital and labor resources necessary for

---

Introduction written by Dimitri UZUNIDIS and Pierre SAULAIS.

carrying out production. The issue of excludability is central to understanding the functioning (creation or development) of a business. Considering, for example, that activities producing scientific and technical information have a larger positive impact (in terms of creating wealth and profit) on a collective level (a large number of businesses) than for an individual business (this is still more frustrating in cases where the business producing information to serve its own innovation activities in fact benefits others), means defending the idea of pooling profits, risks and opportunities. In an economy marked by the asymmetry of information, this does not mean that the selection mechanisms of businesses are unable to operate. It means that external economies, which firms can claim to conquer, are guiding entrepreneurial strategies on localization, competition, spinning off and cooperation.

There is actually a trend for businesses to use their environments to their own advantage rather than to invest, for example, in all stages of technology creation and innovation. This can be explained by the fact that during the acquisition (appropriation) of production resources, all investments are less expensive than those made during resource creation. This, in turn, explains the innovation strategies of organizations and the appeal of a region with plentiful scientific and technical resources. The collective profitability of capital can prove to be greater, while its private profitability can become insufficient. The explanation for the superiority of the social yield from investments in business research and innovation, in comparison to individual capital, lies in the increase in the number of factors that determine the possibility for a given business to make a profit. These overarching factors (education, environment, health, finance, links between industries, communication, needs and aspirations, etc.) have an effect on the marginal cost of a business or activity and, all things being otherwise equal, influence the yield made on the capital invested. The idea of the "network" then emerges as essential to economic observation and analysis. In our economy based on knowledge and open information, the "network" is the industrial structure that brings technical change and guarantees the development of the mode of production and consumption [UZU 12, LIU 16].

The formation of innovation networks follows four paths: the multiplication of exchanges between public research and businesses; the creation of institutions for commercializing research; help for the creation of technology businesses; and the organization of the territory using scientific and technological "centers of excellence". The positive overall business climate ensures the emergence and spread of innovation. An innovative entrepreneur or even an innovative business that acquires scientific and technical knowledge to innovate therefore serves a crucial purpose within a group of interconnections and opportunities for innovation and profit.

To understand and analyze the conception and launch of the innovation process, it is essential to consider the creative dimension of the individual, the business and the organization in general. In new approaches to innovation, the entrepreneur and the business are used according to their skills and their function in the creation of resources; innovation therefore becomes endogenous, gradual or radical, and integrated into a complex process with abundant feedback and interplay. The innovative organization is presented here with a dynamic system made up of specific and varied skills. By acquiring, combining and mobilizing these skills, innovators are able to create technological resources and advance relationships with their environment. This is the reason that managing the conception, application and development is important in the implementation of an innovation process.

The organization (business, institution, spatial innovation system) has a set of knowledge and skills from learning processes, which are integrated into its memory (routines, according to the company's evolutionist approach). The company therefore creates its "knowledge capital" [LAP 16]. The coordination mode and modalities applied here are related to the efficiency and specific features of managing projects and processes. The organization develops and is subject to continuous external aggressions. These aggressions (competition, substitutability of products and technologies, regulation, etc.) are the product of the economic context and have an effect on innovation engines, and also act as a means of selection. Selection procedures are at the heart of the business climate, and include the nature of product markets, the availability of capital and labor, the pace of innovation, the effects of public policies and the protection of intellectual property [LAP 12]. They can, as a result, create alternatives to the mode of functioning, management and production of a given business (or a particular economic system), which is brought about through the market.

The business climate is, in all cases, the creator of either barriers or opportunities for organizations and individuals. The unpredictability of results and the possibility of alternatives are the main uncertainties inherent in innovation activities. On the other hand, just as the innovation process involves learning and development, the innovative organization must constantly make internal adjustments and review its relationships with its technological, economic and social environment.

At the macroeconomic level, innovation requires a continuous reassessment of the social relationships and institutional structures that characterize a given economy at a given time. Innovation is the result of close relationships between producers and consumers, of asymmetry in information or even of microeconomic and macroeconomic growth processes that are out of touch with reality. It is born of imperfection or imbalance, and contributes, in turn, to improving the economic order. It also has a stake in the dynamic growth model based on uncertainty, risk and profit. Innovation is born of uncertainty and risk and, at the same time, creates uncertainty.

The "flaws" that characterize an economic system are, however, a significant source of opportunities for investment, production and the spread of new market values. Yet to arrive at those opportunities, it is necessary for the economic mechanisms to be, at one time or another, in stages relative to each other. Time intervenes in the preparation, organization and, quite simply, the seizure of opportunities that the market offers to agents that are supposed to create new productive combinations.

This book is made up of eight chapters with the aim of discussing the conditions that ensure the implementation of innovation processes, and the launch and distribution of innovations. The two major engines of innovation are knowledge and entrepreneurial spirit [BOU 14]. These engines are initiated by the entrepreneur and the business. The success of the actions of the entrepreneur and the business is the result of norms, rules, traditions and institutions through which economic functions are organized, and thanks to which innovation activities become both individually and collectively profitable. These activities must be considered through the lens of entrepreneurial spirit and creativity, capabilities and strategies for interacting with businesses, consumers, and public authorities, as well as through the means of coordinating their actions with the aim of creating or organizing a market that must recoup investments, alleviate risks and guarantee, for a time, the vitality of business. But how can the entrepreneur use these networks for innovation? What are the factors for success that can turn an idea into an innovation and create value? How can the dissemination of innovation be organized? How can a business transform itself? How can it acquire and enhance the knowledge required for its development? How can it protect its technological and knowledge-related inheritance? How can the economic environment influence innovation engines and systems?

According to Michel Marchesnay, in industrial capitalism, innovation is based on two pillars: understanding technology and having a significant entrepreneurial spirit. Innovation aims to make the capital invested by the entrepreneur profitable. In the first chapter, this link is analyzed at three levels: macroeconomic (nation), mesoeconomic (groups, sectors, regions) and microeconomic (entrepreneurial). The development of industrial capitalism is broken down, according to major inventions, into three generations (1775-1875-1975). Each generation takes place over two stages. The extensive stage focuses on the transfer of inventions to production techniques and tools, while the intensive stage concerns the development of consumer goods and "modern" service activities. Entrepreneurship is thus idealized in the intensive stage as it facilitates the transition to "modernity". However, reality shows that major innovation is dependent on the strategy of industrial and financial groups, innovation policy and the land use policy implemented by the State.

As regards the importance of the territory in which an innovative entrepreneurship emerges, Corinne Tanguy and Dimitri Uzunidis refer to theories of the "innovative milieu". An innovative milieu, defined as a group of actors (businesses, institutions, consumers, etc.) located and settled in a given region, is formed through multilateral socioeconomic interactions. It therefore stimulates the appearance of different kinds of innovation, as well the emergence of new businesses. However, the role of the innovative entrepreneur is crucial to the realization of opportunities and resources offered by an innovative milieu. Indeed, it is this that makes good use of the territory's cognitive, financial or regulatory resources and contributes to the creation of new technologies and jobs through networks and the proximities it is able to make more widespread.

Focusing his analysis on start-ups, Gérard Kokou Akrikpan Dokou shows, in the third chapter, the position of interpersonal relationships in launching an entrepreneurial project. A start-up is an innovative company with a strong foundation in new technologies and good market prospects. An entrepreneur behind such an organization is often focused on the central task of making the most of experience, knowledge and skills. It is therefore necessary for them to have advisers. There are significant requirements for providing such support, as complementary skills are necessary for setting up and making a success of a complex business. In this regard, the logic behind the actions taken by creators is greatly impacted by what they need to be, know and be able to do, to the extent that it fundamentally rebuilds their entrepreneurial identity. Advisers must therefore make these three requirements for a high-tech entrepreneur part of their role as expert counsel in order for the innovative project to succeed.

The skills that are necessary to acquire for innovation are related to entrepreneurial creativity. Creativity and innovation are often addressed together to describe the product-oriented innovation that allows a business to have a "star" product, but this occurs much less often when it comes to studying the successful combinations of multiple ideas that have led to an entrepreneurial success. However, this creative approach can be difficult to emulate, which gives it the protection of a competitive advantage. The aim of Marc Jaillot, in the fourth chapter, is to question the idea of creativity belonging to entrepreneurship in order to determine its place and its importance in small businesses. These are the essential engines of inventiveness, innovation and production in industrial countries.

The question of creativity also arises in large businesses. Maggy Perrier and Audrey Depeige, in the fifth chapter, present the processes and methods used by an international company to create and guarantee value through its innovations. They describe the activities and structure of a business, in terms of contributing to improving performance in innovation and maintaining a competitive advantage. The competitive environment particularly pushes the business to research different forms of collaborating with other organizations in order to create, guarantee and offer value (products or services) by encouraging knowledge-based innovation. In particular, field studies show that the business makes use of two sources for accessing new technologies, turning to networking with other businesses in its sector that are likely to foster the co-development of new products. New forms of partnership here manifest themselves in the emergence of inter-organizational learning and the distribution of new knowledge, supported by dedicated intellectual property strategies.

Yann de Kermadec, in the sixth chapter, explains that at the heart of interactions between innovation, inventions and patents is the language of patents, which becomes very clear to the designer when complicated and often lengthy sentences on patent claims are put into diagrams known as "trees of means". Being both a strategic tool for protecting and using innovations, a mine of information, and a powerful design language, patents allow innovation projects to be energized, secured and used more effectively.

The aim of the seventh chapter, written by Pierre Saulais, is to discuss the connections between invention, innovation and intellectual property rights. Specifically, after having shed light on the three elements mentioned, the author shows that these links are in fact bridges between fields which, until now, have rarely been connected, and that it is none other than knowledge which creates these bridges. General knowledge of objects calls for knowledge of how to become a strong engine of innovation.

However, businesses and, more generally, industry must have an unprecedented resurgence to respond to the challenges presented by a complicated and turbulent present, which is particularly critical from a social, economic and environmental point of view. In the final chapter, Theodor Felezeu analyzes the steps that industry can take in order to become part of a continuous innovation, while also contributing to humankind taking back control of the direction of its development.

In all of these cases, the engines of innovation are formed in three stages: an observation stage, an imagination stage and a persistence stage. Whether formally or implicitly, these three stages also apply as much to an individual creator and entrepreneur as to a business with powerful scientific research structures and engineering technologies.

Bibliography

[BOU 14] BOUTILLIER S., UZUNIDIS D., "The theory of the entrepreneur: from heroic to socialised entrepreneurship", *Journal of Innovation Economics & Management*, vol. 14, no. 2, pp. 9–40, 2014.

[BOU 16] BOUTILLIER S., UZUNIDIS D., *The Entrepreneur*, ISTE Ltd, London and John Wiley & Sons, New York, 2016.

[LAP 12] LAPERCHE B., "Innovation processes: why institutions matter", *Journal of Innovation Economics & Management*, vol. 9, no. 1, pp. 3–11, 2012.

[LAP 16] LAPERCHE B., "Large firms' knowledge capital and innovation networks", *Journal of the Knowledge Economy*, http://link.springer.com/article/10.1007/s13132-016-0391-7, 30 June 2016.

[LIU 16] LIU Z., UZUNIDIS D., "Globalization of R&D, accumulation of knowledge and network innovation: the evolution of the firm's boundaries", *Journal of the Knowledge Economy*, http://link.springer.com/article/10.1007/s13132-016-0381-9, 7 June 2016.

[UZU 12] UZUNIDIS D., BOUTILLIER B., "Globalization of R&D and network innovation: what do we learn from the evolutionist theory?", *Journal of Innovation Economics & Management*, vol. 10, no. 2, pp. 23–52, 2012.

1

Innovation Strategies and Entrepreneurial Dynamics

Humankind has made ceaseless progress in discovering new ways to produce and consume. Its major discoveries have concerned energy sources, means of communication and transport, products and modes of production. They have then spread through migration, exploration and conflict, sometimes over the course of millennia.

However, over barely three centuries, European countries have undergone the "great transformation", to use Karl Polanyi's [POL 44] phrase. In reality, this encompasses three "revolutions", or rather generations, of producing and consuming industrialized goods in line with certain features linked to technical or commercial innovation.

Industrial capitalism is the product of agrarian and mercantile capitalisms, through the search for profit undertaken by entrepreneurial individuals and companies. The industrial entrepreneur appeared first in England, and then in France, in the 18th Century, and became the dominant figure in the following century: the main player in innovation.

Nonetheless, the industrial system follows a threefold interactive process of innovation, internationalization and concentration, as evidenced, for example, by the current "great transformation" in the digital economy. Each generation developed in two stages: the first, focused on major discoveries, developed new stock from productive capital (knowledge, processes, etc.), and the second drew on these innovations to develop products and markets

Chapter written by Michel MARCHESNAY.

further down the line. These new activities spread throughout the world, first to create new areas of production, which in turn become areas of consumption, and then creation, within these new activities. At first regarded as futuristic, they become widespread and commonplace, demanding new forms of innovation focused on methods of consumption.

The following developments organize themselves around the most important questions, with regard to the complex relationship between the dynamism of industrial capitalism and the role of the entrepreneurial spirit. The following issues will be addressed: the nature of entrepreneurial innovation (1), and its level of economic analysis (2). The ideas of industrial generations (3) and then stages (4) are then placed in their historical context. Points 5 and 6 successively address the social question, through the relationship between order and progress, and then the issue of the plurality of the entrepreneur as innovator in the emerging capitalism. Points 7 and 8 consequently address the issue of individual (entrepreneurial) and collective (public) strategies in innovation.

1.1. The entrepreneur: the key player in innovation

An entrepreneur is, from a legal and economic standpoint, someone who uses their own capital (financial, personal, institutional, etc.) to take the risk of creating (or taking on) a business, with the aim of drawing various returns on investment from it (revenue, personal and social satisfaction, etc.) [JUL 98]. The desired result is as significant as the risk taken, and, in particular, it is as great as the degree of innovation. However, the range of expectations increases the complexity of innovation practices.

1.1.1. *By their very nature, every entrepreneur takes a risk dependent on innovation*

From the moment that an entrepreneur creates and manages his organization, he is inevitably met with uncertainty. Moreover, they contribute ever more to increasing that uncertainty as they are lead, whether voluntarily or not, to innovate. Whether they choose to adopt a strategy based on leading or on following, by intent or by adaptation, this operating logic has its economic rationale in expectation of a surplus, a return in an economic sense, related to appeal for clients, on an unstable market open to opportunities. Contrarily, and theoretically, the less the entrepreneur is an innovator and a risk-taker, the less he can hope to make their investment profitable [LOW 70]. However, this logical explanation does not hold up when:

– the market is overprotected, in which case it creates subsidies, hardly encouraging innovation;

– the market is too easily accessible, in which case the cost of innovation is not in line with the risks of failure or hypercompetition;

– the market is dependent on the sector in such a way that the profits of innovation and surplus factors are picked up by partners before or after, which can dissuade people from undertaking unprotected innovations.

1.1.2. *Innovation restores a range of logical structures and practices in entrepreneurial action*

Individual and emerging entrepreneurial innovation must be characterized by managerial, organic and planned (deliberate) innovation. Each story of innovation is unique as it depends largely on the innovator: who they are, what they want, what they know and what they do. It is then dependent on internal social interaction (the culture of the organization), knowledge and know-how (working experience), and meeting the pressing and underlying needs of the market (the mission). In reality, all of these data are variable and constantly changing.

Nonetheless, there are several points shared in all innovative approaches. The most common is for the innovative entrepreneur to draw on experience acquired through education (*learning*) or through training (*doing*), be they a pastry chef or a doctor or engineer. More broadly, the innovator enters into an atmosphere favorable to their operating logic, whether it be a profession, an environment or a community. In terms of legitimacy, their expectations will be at once economic (the feasibility and reception of the product) and social (recognition, from peers as much as from social networks, both local and further afield).

1.2. Industrial analysis of innovation

Industrial analysis addresses the issue of innovation from two distinct angles: the interaction between structures, conduct and performance (SCP model), or the differentiation of three areas of study – the sector, the group and the region.

1.2.1. *The analytical approach: the SCP model*

Industry is defined as a competitive field of companies adopting "conducts" (behaviors) ranging from conflict, collusion, cooperation. Each industry defines itself empirically according to its structures, including its degree of concentration, the elimination of potential competitors, etc. Lastly, it is characterized by its performance, particularly in profitability and growth. In these three areas, industrial innovation strategies wield a dominant influence. In particular, certain innovations are likely to have a destructive and reconstructive effect in a structure–strategy pairing, owing to technological breakthroughs (processes) or commercial ones (markets). In this way, the areas of strategic activity are continually disturbed by innovations, whether they arise from established companies or contenders just starting out.

1.2.2. *The institutional approach: mesosystems*

The intermediary approach between the microeconomic and macroeconomic levels provides a conceptual and empirical framework that allows links to be made between operational logic and practices. As regards innovation, it specifically offers a framework for evaluating and promoting the policies of companies (groups or SMEs) and public institutions.

– The *sector*, as an object of research, refers to the series of steps in a transformation up to the final stage. Depending on the case in question, it breaks itself down (from top to bottom) or rebuilds itself. The sector is constantly changing, mainly at the whim of innovations in one of its areas. In the example of innovations in digital technology, broadly speaking, the overall economy of the sector can be continually questioned. In particular, the operational actions of the players develop, bringing with them structural changes, adaptation strategies, or even breakdowns in the face of innovation, and, as a result, variations in performance.

– The *group* brings together companies controlled, whether legally or not, by a core group of owners. The prevailing managerial orthodoxy dominates in this area, with the aim of optimizing the capital invested. In theory, the group is, according to a well-known adage, "an island of conscious power in an ocean of unconscious coordination (the market)". As a result, it has an independent strategy towards innovation which materializes in the form of an integrated technoscience system. By relying on its own laboratories, the executive ranks plan the pace and type of innovations, most often through technology transfers. However, some innovations rise unpredictably, uncontrollably or beyond the scope of the group's laboratories. This issue arises most often at the two ends of the process: either for product-processes at the discovery stage, or for marketable products. In these two cases, small businesses, according to entrepreneurial logic, have proven to be more capable of withstanding risks: a case in point is pharmaceutical groups, which are continuously looking for patents or licenses from start-ups.

– The approach in terms of *region* has sparked a vast amount of research, including from geographers and historians, into innovation structures and policies. Alfred Marshall, in Cambridge (England) at the end of the 19th Century, evoked the idea of ethos under the term "atmosphere". In this way, certain geographic areas, both urban and rural, known as "districts" (of wool, cotton, cutlery, etc.), made up preferred locations for promoting, encouraging and assimilating innovations in a given sector. Nowadays, Marshall's districts have been expanded and are known as "clusters", with multiple forms of agglomerations of businesses, laboratories, etc., whether in a given sector (technology park) focused on one industry, or in a specific urban area (technology hub), such as a network of hotspots and research laboratories on various subjects.

1.3. The three eras of industrial capitalism

1.3.1. *The origins of industrial capitalism*

Industrial capitalism is the result of a series of major innovations, or breakthroughs, described by Adam Smith upon their inception (1776) and theorized in around 1900 by Schumpeter (using the cluster innovations of the second industrial generation). This type of capitalism is different from agrarian and mercantile capitalisms due to the formation, with the aid of financial capital, of a set of physical and intellectual assets that make up what is known as the "catalog of technologies" (*blue book*) and is available at all times. The capital is invested by entrepreneurs and their agents in the development of mechanization, the division of labor, and the physical and legal consolidation in companies (manufacturing). The task falling to entrepreneurs requires a "talent for administration", according to the French economist Jean-Baptiste Say, who was himself a manufacturer in France during the First Empire.

The development of industrial capitalism has historically occurred alongside three underlying trends related to researching new opportunities and new prospects: this might be, on the one hand, the continuous, and if possible controlled, production of innovations; a focus as the markets reached maturity, on the other; or finally internationalization, in search of new prospects. The globalization of industrial capitalism was strengthened by the fact that major innovations – breakthroughs – were introduced in new countries.

1.3.2. *The development of industrial capitalism: the concept of generations*

Generations can be said to start in line with the emergence of their products and services in around 1780 (steam engines, weaving machines, spinning machines), 1880 (steel industry, internal combustion engines, electricity, etc.) and 1980 (computer technology, biotechnologies, etc.). For each industrial generation, there is an almost identical mode of development, indicated by an innovation process moving from basic research, to applied research, to the development of technically viable and commercially appealing products [JUL 97]. In a way, it moves from a scientific basis to a technological one, which generates new production processes. This leads to "machines" (intermediary goods and equipment) that enable the production of goods and services targeting the mass market, before turning to increasingly specialized customer groups, by relying on services responsible for creating and distributing ever more advanced innovations.

Use of the term "*industrial revolution*" to describe the first generation is misleading. It in fact brings together activities carried out over almost three quarters of a century (1710–1780) to design a tentative series of aims (like the steam engine) and, primarily, experiments (weaving and spinning machines) that lead to a technically operational invention following many mistakes, failures and collapses, as described in the works of Paul Mantoux, David Landes [LAN 69, LAN 98] and Denis Woronof [WOR 94]. Only innovative entrepreneurs with recognized business skills (*risk takers, organization managers, market builders*) eke out a living, but they must also have luck, reminiscent of the natural selection process in Darwinism. This period is one of a "savage capitalism", where everything is permitted (such as "appropriating" the basic principle behind an invention or copying the development processes of competitors).

While this same initial (extensive) period of trial and error exists in the second industrial generation (the end of the 19th Century) during the move towards new TPM (technologies, products, markets) groups, the fundamental research is from this point onwards restricted to public and private laboratories and the development of new product markets through the management of industrial and financial groups. This is a technoscience system (the Pasteur Institute is an example of this dual affiliation).

The third industrial revolution, based on the ongoing digital revolution in particular, is, due to the complexity and growing costs of innovation, giving rise to networks of alliances, involving ever longer periods of time between design and marketing, and strategies mixing competition and cooperation ("coopetition") depending on the stage of the industrial innovation process.

1.4. Extensive and intensive stages

The French school, known as the "regulation school", supposes that each industrial generation has historically had two successive stages, the extensive and the intensive.

1.4.1. *The extensive stage*

This centers on the increased accumulation (*widening*) of a productive capital (of material and immaterial goods and services) intended to design and create a large amount of innovative productive capital. For example, the steam engine is the product of both innovation in the steel industry (Abraham Darby's coke furnace, which replaced wood, at the beginning of the century) and research in physics (the Carnot principles). The replacement of wooden rails with iron rails stimulated the creation of railway networks (1830). Subsequently, the second industrial generation (1880), born in new countries (Germany, the USA), adopted the breakthrough innovation of the open hearth furnace, which was intended to increase the durability of steel rails.

There are many more examples of cases showing intergenerational continuity in the flow of innovations. In this way, without going back to Volta, inventor of the electric battery in around 1800, the "innovation clusters" in the electrical industry (1879, electric bulb) find a new development in the early computer, ENIAC, dating back to 1942. Research on materials, particularly silicon, led to the electronic chip in 1975. Since then, debate has raged around a design, whether centralized by super-computers, or decentralized (shared) around micro-computing (1980).

1.4.2. *The intensive stage*

The second stage is called "intensive" as the technological capital is sufficiently stable for controlled usage to begin in researching productivity and an increased range of techno-scientific applications. It consists of developing the sources of applications and diversifying the range of products (goods and services) intended for eventual consumption (businesses, then individuals). The innovations have the strategic aim of broadening the range of target markets (without actually creating them) or deepening the supply within a range. It therefore revolves around provoking new needs or improving the satisfaction of existing needs. In other words, in the intensive stage, company growth is boosted by commercial use of innovations based on major discoveries: a "spillover effect". An innovating company that has patented its design innovations benefits from the bonus of being the "first arrival" (who will establish the rules), and can go on to make investments in development. An example is IBM in the 1970s and 1980s which, knowing that the PC revolution would soon explode, in both the supply and design of digital items, turned to providing services, realizing that existing industrial groups, at the end of a concentration process, would become obliged to do so.

1.5. Innovation: between order and progress

The Brazilian motto (*Ordem e Progresso*) reflects Schumpeter's idea of "creative destruction", as an industrial society bases itself on the ideology of progress. Maintenance of the existing order is accused of creating situations that generate private profit on the basis of past innovations. Just like the Titan Prometheus, it belongs to the rising classes, who "have everything to gain and nothing to lose" by launching themselves into adventure (see the "*venturer*", a North American entrepreneur) and innovation, stirring unrest – particularly, the many dominant activities of the previous generation. Nonetheless, in the intensive stage, industrial society favors social order based on a middle class that consumes its own goods and services, and is open to modernity.

1.5.1. *The role of ideologies*

1) The first generation is based on the ideology of the Enlightenment. This advocates individual ownership (the French word "liberté", or "freedom", is the most used word in the Declaration of the Rights of Man). The rising class of manufacturers, often emerging from modest parts of a less developed State, acquire the status of petty bourgeoisie, by arranging the mechanization and division of labor in their factories (often producing national goods). However, technical progress occurred slowly in France in comparison to England; home-based work and hydraulic energy, to take two examples, lasted until the end of the century.

Moreover, the majority of great industrial dynasties (the "conquering bourgeois") see their heirs adopting positions of private income by living on the gains (patents, brands, etc.) of the "founding father". Thorstein Veblen notably described this "leisure class" in the USA in the 1910s [VEB 99].

2) The second industrial (and social) generation was based on a scientific and rationalist ideology. Leading to the "age of technologies", it took its inspiration from, among others, the positivism of Auguste Comte, the industrialism of Saint-Simon, the utilitarianism of J-S Mill and the pragmatism of W. James. Innovation was the product of a deliberate and scientific practice enabling it to move from manufacturing to laboratories, be they public (universities, engineering schools) or private, such as the Pasteur Institute in France.

However, this ideology of scientific progress, prioritizing the efficiency of the productivity–profitability pairing, would largely develop after 1870 in new countries: Germany and the USA. It was these countries that would establish a rational and pragmatic relationship aimed at effective action between the research (basic and applied) and development of new products and procedures, by means of technology transfers. This approach would be systematized in the scientific organization of labor (Taylor) and the science of business administration (Fayol). Thorstein Veblen, in the 1900s, recommended that the USA leave design and decision-making to engineers (later to managers), rather than to financiers, and especially not to "venturers" who behaved in a predatory manner.

The *intensive stage* began in the USA in the 1920s. Many large brands of consumer products date back to this period. They were developed, designed and "marketed" in laboratories and company services. However, the "*affluent society*", supported by the "*staff and line*" structure of companies, did not emerge until after 1945 in the USA, and after 1965 in Europe. After 1975, there was a period of slowing down, followed by stagnation and even decline in mass products, drawing concentration in many sectors to the second generation. At the same time, the consumer society reached new countries, particularly in the Pacific region, followed by the BRIC countries (Brazil, Russia, India, China), leading companies to think globally in their approaches to innovation.

3) As regards the third generation, dating it has become particularly complicated as it has brought about significant transformations in the sectors of preceding generations – in the same way, for example, that electricity and the internal combustion engine, among other innovations, turned the textile industry upside down. As mentioned above, since the 1940s, fundamental (atoms, robotics) and exploratory (new materials) research have heralded future innovations. 1975 was a pivotal year: the consumer society was under siege (saturation was widespread and opportunities were drying up), and the industrial concentration had turned into closures and unemployment, to the extent of referring to a "post-industrial" society, based on innovations in services and SMEs, and marked by "the return of the entrepreneur" (as opposed to the manager). In fact, the restructuring of the sectors of the two previous generations has continued, particularly in computerization and digital technology. The extensive stage, of designing transferable procedures and materials in terms of technology, is considered to have started in the 1990s with, among others, the transition towards the use of personal computers. This has been marked by the arrival of new entrepreneurs who have become global icons, and who will, in turn, take the places of Rockefeller, J.P. Morgan, Edison, etc., in the gallery of Schumpeterian entrepreneurs.

1.5.2. *The role of the "zeitgeist", the spirit of the age*

As in the case of the two preceding generations, there can be no industrial revolution if the mind is not ideologically ready: take, for example, the Enlightenment thinkers in the 18th Century, or representative democracy in the 19th Century. The destruction related to breakthrough innovations has stirred up conflict and debate on the nature of the new society; it is thus obvious that new technologies in information production and consumption, based on the digital economy and society, call into question the components of a democratic society. This debate is also related to the continuation of the long threefold movement in industrial society, which will be international going forward, and is continually innovative, even as economic and financial power is concentrated in a decreasing number of individuals and institutions.

For their part, the "progressive" intellectual elites will contribute to the widespread acceptance of innovations and, more broadly, ideological breaks and social transformations, causing a resurgence of the middle classes. In the 18th Century, this was the role of encyclopedists. From 1835, the daily press began to develop, with writers of serials (Balzac, Dickens, etc.) describing the positive and negative effects of industrialization by pitting them against traditional structures. In so doing, they kindled the demand for new products. The consumer society, in the following century, would be largely reliant on advertising spread by the media (radio, then television), to the extent that J. K Galbraith spoke of an inverted sector (in which the producer dictates the preferences of the consumer) and Vance Packard of "hidden persuasion".

In the 21st Century, the digital revolution has multiplied the distribution channels. They have both an almost complete global audience, and a direct relationship with potential consumers. This has been followed by an intensification in the logic of innovation, which manifests itself in a reduction in the lifetime of each product, and at the same time an expectation for it to be replaced by a "better" product. Such operational innovation logic is based on social practices, particularly those of the "elites" thought to be leading opinion-makers and to ensure "visibility" (as well as intelligibility for technology products).

1.6. Innovation and the "technical age": a Pandora's box?

It should be noted that this question is not a new one. Let us recall the myth of the Titan, Prometheus. He was punished for eternity for having given fire to humankind, and particularly for having let his brother Epimetheus, seduced by the nymph Pandora, open the box (filled with knowledge) from which all the evils of the Earth escaped. In other words, Promethean innovation has neither a constructive nor a destructive nature with regard to the actions of societies: many innovations that are beneficial in times of peace were developed in war. This is also the case for the digital revolution, which may or may not foreshadow an Orwellian world, depending on its usage. Such thoughts lead us to an old question: is the entrepreneur Prometheus or Epimetheus?

1.6.1. *The innovative entrepreneur: is there an ideal type?*

Over the course of these three generations, common features of the *ideal type of innovative entrepreneur* have been made clear. The majority of these entrepreneurs emerge from the middle class and are motivated by a desire for upward social mobility. These "*self-made men*" have shown themselves to be opportunists and fine strategists. While controlling the process of designing and then making use of the innovation, they work to retain intellectual property to the extent of flirting with the boundaries of legality, at the same time as overseeing the multiple other processes than often lie in their path.

Indeed, the "traditional" starting point of many creators allows them to acquire, whether consciously or not, certain pragmatic capabilities. They are therefore able to concentrate on problems rather than preconceived solutions; they are good at comparing their operating logic and their practices, and at complying with an experimental approach of trial and error, to the point of modifying, or even completely changing the design model to alter, explain and even justify the approach.

This mindset and behavior is central to innovation approaches inspired by the pragmatist John Dewey. Based on a "*trial and error process*", the development of, *inter alia*, learning processes, training courses, alternations and case processing is part of a political will to promote the creation of innovative, and also lasting, businesses, including through the acquisition of unique, experience-based "know-how".

Schumpeter, on the other hand, observed that the "innovation" stage is followed by a consolidation and profitability stage, in order to lead to a desire for its continuation by building a *dynasty*, that is, a kind of family-based governance, sometimes lasting for centuries. However, this ideal type of "dynastic entrepreneur" (and multigenerational entrepreneurs) must not ignore the crucial role of millions of small- and medium-sized businesses which, in the extensive and intensive stages, contribute, through their productivity (in terms of both creation and creativity), to strengthening the development of the constituent procedures and processes of sectors in the extensive stage, and of the downstream markets in the intensive stage.

Moreover, the entrepreneurial spirit is not limited to individual entrepreneurs, owner-managers of their own, duly registered, businesses. Indeed, it is noticeable that some entrepreneurs have a weak entrepreneurial spirit (unadventurous, poorly organized, not creative, with low incomes), while, increasingly in the emerging society, many people or institutions with public or social purposes show a strong entrepreneurial spirit. We have seen high-ranking State civil servants, both civilian and military, and "visible" individuals who have implemented highly innovative humanitarian projects. In this regard, social entrepreneurs play a growing role in replacing inadequate provisions.

The "digital revolution" has contributed to the exponential development of networking, leading to new forms of collaboration which abound with opportunities for innovation. An example is the development of the "collaborative economy" to its, sometimes blurry and often changing, boundaries between charitable work and making profit. Incidentally, groups, and companies more broadly, draw and retrieve innovative ideas from there, if only to withstand innovations in service provision.

1.6.2. *Types of innovative entrepreneur*

1) The typical entrepreneur develops an individual project by making use of various resources with the aim of making a profit, residual income and randomized income (an uncertain "gage", to use the expression suggested by Richard Cantillon in the 18th Century). This profit-making objective varies depending on the character, skills and strategy of the entrepreneur showing an inclination for innovation. The expected benefit therefore goes from a mere subsistence income for a person looking for a job, to expectation of a "*jackpot*" (for example, a group buying out the entrepreneur's business) for a start-up.

Depending on the type of entrepreneur, the innovation can appear for highly variable reasons and according to practices of very different levels of intensity. A "high", and broadly managerial, definition has long been retained from the "*Frascati manual*", in which innovation is the product of a deliberate approach that leads to strategic consequences for the economic model (technologies, products, markets), and which particularly manifests itself in the form of filing patents, licenses and trademarks.

However, it is telling that the definition of "innovative" has subsequently expanded progressively [DRU 84]. In particular, various kinds of changes carried out in the organization of businesses at the so-called "human level" have been retained. We can therefore say that, broadly speaking, there is "innovation" from the moment that the change carried out (for example, a new machine, or a new procedure or process, or a redesign of a product or even of a target market) corresponds to the entrepreneur's operating logic, after which he or she is considered "innovative", albeit sometimes only in hindsight.

2) We can outline a categorization of kinds of innovation, as perceived by entrepreneurs, particularly in small- and medium-sized businesses.

First, there is the pioneering innovator, who designs and/or markets a product (or service) born of his or her creativity. This is related not only to his or her experience and/or education (his or her training), but also to his or her social capital (particularly, the community to which they belong) and his or her personal tastes. Take, for example, start-ups with skills (knowledge and know-how) in a specialized area, such as certain types of video game (recreational, educational, etc.).

Next comes the imitator (or follower), who wants to establish himself or herself in an emerging activity or one undergoing reorganization. They could draw on experience acquired during the initial stage, as well as from the learning and from the specific manners of innovative goods and services. They may, for example, move into an existing type of video game.

The adaptor is an entrepreneur who is already established, but who modifies their economic model. This change in direction generally requires new resources and skills within the framework of the existing business. In this way, the adaptor is able to refocus on, for example, games intended for a subset of players.

Lastly, the convert looks for another activity, most often in another location. Some "*serial entrepreneurs*" change their activities at the whim of emerging opportunities, particularly in the retail trade, as evidenced by frequent changes in ownership, but also in trade in very robust areas. In this way, the convert can move from video games to another activity in the digital sector, or even outside it; this "nomadic" practice can be found in activities with short life cycles. Examples include fast foods, tattooing, electronic cigarettes and "green" products.

1.6.3. *The capabilities of the innovative entrepreneur*

1) Although there is a wide variety of entrepreneurs, there are still traits unique to the innovative entrepreneur ("enterprising"). In addition to traits relating to character and traditional professional capabilities, the entrepreneur's "occupation" includes psychosocial skills related to their community and personal history. Entrepreneurs, or at least the most innovative ones, most often have a personality notable for its acceptance of the risks of business and opportunism.

Within the business itself, the issue of authority can arise as soon as the entrepreneur surrounds himself or herself with collaborators (business partners, associates, employees, etc.) during the implementation of innovative projects, thereby creating what an economist (Nobel Prize winner Oliver Williamson) has called the risk of opportunism. Factors including resistance to change, or even incompetence or, worse, ill-intent, can arise from various actors, including external stakeholders (those providing funding or giving orders). On the other hand, an entrepreneur with a successful business can be tempted to innovate without realizing all of the harmful consequences, particularly the risk of losing independence.

2) The management of innovation, in both large businesses and group subsidiaries, is based on a hierarchized and rationalized procedure which is formalized in a plan spanning several years. SMBs and middle market enterprises have varied reasoning and practices according to their power and market position. Conversely, all small businesses led by an owner-manager are largely governed according to an emerging process based on the perception of innovation *opportunities*, and its presentation of implementation ("feasibility"), suitability of its skills and markets ("consistency") and the expected consequence for the business (in the event of either success or failure). Small businesses are in fact particularly aware of two major strategic risks – vulnerability and dependence – in a particular manner that determines their position in the extensive or intensive stage of an industrial generation.

1.7. The major strategic choices

1.7.1. *Breakthrough innovation or adaptive development?*

The initial, extensive stage is characterized by intensive innovative activity, coming largely from very small innovative businesses managed by "pioneers". Within this forming industrial structure, innovation primarily focuses on intermediate goods (raw materials) and equipment (machines, means of communication, infrastructure). The entry barriers on emerging markets are low, encouraging the growth of small-scale entrepreneurs looking for reliable procedures and processes: for example, the spinning jenny and the water frame (1730–1780), electricity, artificial dyes, the combustion engine (1880), the micro-computer and the laser (1980). We are used to mentioning the, often mythologized, cases of great innovations that have been (or could be) developed in workshop or a garage. The fact that some innovations have serendipitously been made following an error in handling (vacuum cleaner) or reinterpretation (penicillin) adds to the tumultuous nature of this stage, and therefore, to the vulnerability of young start-ups, recognizing that the success of some leads to a selection process that is difficult to manage.

Generally speaking, innovations, as basic principles, have stabilized (as in the case of information technology) and come into the public domain, and have focused on adaptations and developments to professional applications, then to private usage. This accessibility will also reinforce the turbulent nature of the market, and increase competition, limiting innovators to increasingly specialized applications and creating a monopolistic situation (transitory, without specific protections) in niches, micromarkets and even nanomarkets. Innovations will therefore affect applications heading for final demand. The "overflow" to consumer markets for goods and services contributes to multiplying innovations in new markets, for example, women's fashion and journalism after 1835, and the car and the television after 1935.

1.7.2. *Dependence or individuality? ("One of a kind")*

1) In this way, we have witnessed a fairly brutal exclusion of managerial SMBs and very small-scale businesses, particularly during free trade agreements (1815, 1860, 1960), for want of being able to take part in innovative sectors. More generally, let us recall that, for three generations, there has been a trend towards internationalization of the markets of the previous generation when they reach maturity, or even saturation, in parallel with a trend towards concentration. In fact, industrial groups involved in business and finance have endeavored to manage and plan innovation, as Schumpeter predicted.

The consequence for entrepreneurial and/or family businesses was a risk of dependency on these, often international, groups. This dependency effect manifests itself when the number of (effective or potential) partners is low (able to become a monopoly), when it is difficult, or even impossible, to replace them with a competitor, and, finally, when this relationship is vital to the business. However, some businesses, particularly in the service sector, recognize this dependency relationship, especially for subcontractors and franchisees. In the digital revolution, dependency reveals itself through submission to both technical and administrative standards imposed by groups, particularly to be referenced.

2) Entrepreneurs then seek to reduce these dependency risks through innovative practices, when the technical and competitive conditions allow it, in terms of production costs and market access:

– they can look for an opportunity for a competitive advantage based on proximity, be it geographic, such as innovation of local products, or commercial, such as a network of social relationships with clients and providers, and more generally with various other stakeholders in the region (particularly institutions supporting entrepreneurship). In this way, small local businesses often demonstrate greater flexibility than the subsidiaries or establishments attached to large groups;

– but, in the face of the ultra-segmentation of markets and increasing turbulence in practices and type of consumption, the opportunities for innovative projects have continued to multiply, justifying the increasing creation of small businesses based on a unique – one-of-a-kind – activity. In the best cases, this relies on skills that are difficult to access and teach, and which cannot be transferred (to another business, for example), requiring a specific "capital" (technical, intellectual, social, etc.).

The uniqueness strategy therefore demands reliance on a permanent *creativity*, a bonanza difficult to achieve in groups governed according to managerial standards. It is likely to immediately lead to extra-territorial, or even global, markets (references are made to businesses *born global*). In this case of an "interstitial monopoly", proximity is no longer geographic, but technological and centered around assembled knowledge facing a network made up of competitors who are both adversaries and allies in promoting innovative activity.

In this "*uniqueness economy*" [KAR 07], the *unique skills* come both from individuals, as in the case of art forms, and "teams", such as in hospitality and high-class catering. Innovation and creativity are reliant on individual knowledge, as evidenced by micro-entrepreneurs in show business, technicians and actors working for themselves (occasional workers). This uniqueness is increased in the *net economy*, which promotes and accelerates the ever faster distribution of innovative projects which have an important singularity. This form of downstream "overflow" from basic industries not only concerns services, but also affects upstream sectors, right up to primary manufacturing sectors. Take, for example, the explosion of organic or locally grown produce, bolstered by the shortening of distribution chains and by direct sales, including on the Internet.

In any case, it should be recalled that the most important thing about "entrepreneurial" innovations, therefore those made a reality by entrepreneurs and their entourages, is, in the overwhelming majority of cases, the changes and variations in the composition and offer of products (goods and services) which appear marginal, but often represent a major change in the strategy (project, economic model, vision, etc.) of the entrepreneur and their entourage (family, colleagues). This observation is in line with the prevailing opinion, which states that entrepreneurial innovations are mainly emergent (and non-radical) and "incremental", appearing through successive additions. In cases of radical change, the entrepreneur changes the name of his or her company, or the business is taken over by a buyer.

1.8. The spread of innovation

1.8.1. *The transfer: the basis of national technological policy*

The proponents of liberalism maintain that innovations are created when underlying opportunities are taken by individuals with an aptitude for entrepreneurship, which presupposes that there is a sufficient "market" of entrepreneurial people. This has historically been the case when existing productive and social structures undergo a significant crisis, threatening jobs and societal positions. It is therefore necessary for competition to be as open as possible (the theory presented in the Treaty on European Union, otherwise known as the Maastricht Treaty).

This approach is said to be Darwinian, as it is based on selection of the most able, or of the luckiest. Followers of the planning school, who advocate a projection-based approach in designing and developing innovations, consider the approach wasteful in terms of resources. The choices are the result of priorities, depending on technological excellence, economic competitivity or social well-being. They are formalized through multi-year programs which determine the goals to be reached and the means to be used.

Countries most often adopt indicative planning based on economic studies in order to give an advantage to the programs considered to be priorities. These programs, which focus on intermediate levels (mesosystems), have primarily concerned the role of groups, regions and subsidiaries as generators of industrial policy, particularly in the area of research and development. Going forward, the digital revolution will lend networks great importance by emphasizing the interaction of individual and institutional initiatives in the conception (research) and realization (development) of innovations in every industrial mesosystem (broadly speaking).

1.8.2. *The subsidiary: an instrument of the industrial strategy of groups*

1) An *industry* is made up of a set of actors (businesses and other institutions) and activities that are connected, whether in terms of competition or cooperation, in such a way that, as a system, the strategy of each company has consequences for other players in the industry. In reality, each *industrial system* has its own structures and boundaries which develop at the mercy of technical and structural innovations and transformations.

The nuclear generator industry is an example of the multiplicity of such entry barriers. In an industry that has reached maturity, such as the automobile industry, the actors are well defined, the barriers to entry are high, and the process of concentration and internationalization continues, while the innovations have more to do with adaptation (proliferation and rotation of models) than with ruptures, as evidenced by the resistance to new processes (recalling the first cars that ran on water or electricity). Conversely, in "young" industries, the barriers are still low and entry is relatively easy, as seen, for example, in the digital industry. The inherent risks of innovation are especially high in a "young" and highly turbulent area where the entry barriers are low and the risks of imitation are high.

2) As an industry enters the development stage, there is a move towards concentration justified by the research of so-called economies "of scale". These include cost savings obtained through improvements, and technology transfers acquired through research laboratories (both public and private). This race towards economies of scale prompts agreements, or even mergers, between companies in industries with compatible activities. It also encourages them to develop internationally by seeking agreements with national companies.

Nonetheless, beyond the advantages linked to increased productivity, brought about by "innovative" improvements, the growth in the profitability of groups is primarily the outcome of transferring surplus productivity achieved by dependent companies, for which the implementation of technical or commercial innovations has been reserved.

1.8.3. *The region: a hub of technological development*

The concept of an innovative environment comes from industrial history, indeed from the history of capitalism, and includes agricultural, mercantile and financial sectors. The first industrial generation was largely based on primary local industries (wool, natural dyes, smithery, etc.), which had sometimes been operating for centuries. The second industrial generation centers more on transformation activities (steelworks, "artificial" chemistry, electricity, etc.) following discoveries made by laboratories or of individual innovations. It opened the way to extra-territorial markets thanks to new means of communication (transportation, telecommunications, etc.), colonialism, and the emergence of newly industrialized countries.

It is this combination of physical and technical data which, alongside social and human factors, characterizes the industrial atmosphere of a "country" (in a geographic sense) or a region. In a best case scenario, the territory acquires a reputation for its notable skills in an activity, which are the result of knowledge (sometimes passed down from ancestors), know-how (sometimes rustic or highly idiosyncratic) or common values (grouped together under the ambiguous term "soft skills"). Since the 19th Century, English geographers have used the term "industrial district" to define a region that specializes in an industry, such as the English wool, cotton, cutlery and pottery districts, among others. Within each of those, companies direct innovations and adaptations by relying on a platform of shared knowledge. This concept has recently been taken up again, using the works of Alfred Marshall, to explain how, for centuries, Italian districts have practiced a "flexible specialization" policy, with each business having a distinctive skill in the textile production chain, as in the Prato region of Italy, for example.

In that spirit, the concept of the *cluster* was developed in the USA. This neologism emerged from the transition, from the 1980s onwards, from the extensive stage to the intensive stage of the digital and biotechnological revolutions. Each cluster has several hubs: a growing town (with amenities and access), a university area (with laboratories), welcoming conditions particularly adapted to the arrival of start-ups in *nurseries* and *Business Innovation Centers*. In contrast to the Italian districts, clusters have broader specializations, the network of companies is less hierarchical and the topics (programs) of research are more varied. Moreover, the networks are at once progressive and outreach-oriented, to the extent that these companies are very often involved in business with the integrated or quasi-integrated area of multinational firms (such start-ups are, by their nature, *born global*).

1.9. Conclusion

One conclusion is clear: since *Homo habilis* began to make tools (no doubt by imitating large apes), Humanity has not ceased to create and adopt *techne* before adapting it to its purposes, as much for strenuous domestic activities as for polite ones. However, industrial capitalism has granted an increasing role to technology, leading to questions of its "utility", philosophically speaking (in the works of Heidegger, for instance). Finally, the transfer of technology currently affects and questions almost all disciplines, primarily those in science and technology, but also those concerning humankind and society, as evidenced by the debates surrounding sustainable development, the modalities of technology transfers and their impact, both in the business world and on their economic and social environments.

The current era is one of breaking the system, of a new "great transformation"; globalization has multiplied the levels of industrial development globally, through the transfers of production methods and technology to less developed countries. At the same time, while the world economy is dividing, moving from a hierarchized structure to an interconnected system through digital networks, advances in cognitive sciences are emphasizing the similarity with neuron networks, raising the key question of human creativity once again. Previously addressed by Herbert Simon, the answer is simple: technological tools function according to enlarged rationality, and humans according to an expanded rationality.

It has become almost mundane to cite the "digital revolution" and all its impacts, not only on technology, but also on individual behaviors and social values. In particular, new technologies have contributed to the development of what is known as hypermodernity. The upcoming generations, the breeding ground for the middle classes, are "network individualists", at once numerous and nomadic. This manifests itself in a desire, be it accomplished or not, to be their own boss, and to stand out by creating their own business. The fourth estate, the media, eases such fears of self and of "visibility" by highlighting the "innovators" and the "creatives", while emphasizing the "deceptions of management", and even "managerial horrors". One of the significant points of the first decade of the 21st Century has therefore been the rise in the legitimacy of the entrepreneurial innovation. Admittedly, the spotlight on start-ups and other innovative businesses provided by the media risks masking the dark side of entrepreneurship, namely those who are entrepreneurs "by force", excluded from paid employment in a former industrial society. They are nonetheless rebuked both for the "cult of performance" and "urgency", to the detriment of innovation and creativity, and are accused on calling the existing order into question.

Accordingly, the myth of Prometheus has not ceased to enthrall us. While the innovations of capitalism have continued to improve the human condition, we cannot ignore the damage caused by progress. This is without question that for which the Gods would rebuke the "man in revolt", to use the expression of French philosopher Albert Camus.

1.10. Bibliography

[DRU 84] DRUCKER P., *Innovation and Entrepreneurship: Practice and Principles*, HarperCollins, London, 1984.

[JUL 97] JULIEN P.A., MARCHESNAY M., *Economie et stratégie industrielles*, Economica, Paris, 1997.

[JUL 98] JULIEN P.A. (ed.), *The State of the Art in Small Business and Entrepreneurhip*, Ashgate, Farnham, 1998.

[KAR 07] KARPIK L., *L'économie des singularités*, Gallimard, Paris, 2007.

[LAN 69] LANDES D.S., *The Prometheus Unbound*, Oxford University Press, London, 1969.

[LAN 98] LANDES D.S., *The Wealth and Poverty of Nations*, Abacus, London, 1998.

[LOW 70] LOW R.E., *Modern Economic Organization*, R. Irwin, Homewood, 1970.

[POL 44] POLANYI K., *The Great Transformation*, Beacon Press, Boston, 1944.

[RIF 11] RIFKIN J., *The Third Industrial Revolution*, Palgrave Macmillan, Basingstoke, 2011.

[SHO 65] SHONFIELD A., *Modern Capitalism: The Changing Balance of Public and Private Power*, Oxford University Press, London, 1965.

[STO 82] STOREY D.J., *Entrepreneurship and the New Firm*, Croom, Helm, London, 1982.

[VEB 99] VEBLEN T., *The Theory of the Leisure Class*, Macmillan, London, 1899.

[WOR 94] WORONOF D., *Histoire de l'industrie en France*, Seuil, Paris, 1994.

2

Innovative Milieus and Innovative Entrepreneurship

Technological innovation is the most reliable means for rebuilding, transforming and expanding markets. The growth of companies and the emphatic globalization of markets have shown the importance of local pockets of productive resources. The geographic proximity of science, technology, industry and finance contributes to the emergence of innovations. Interactions are organized through the combined interplay of private actors and public institutions. While territory has been considered an adjustment variable with no independent existence in economics for a very long time, economists are currently leaning towards the "local economy" as the geographic and economic level of production organization, and, consequently, of the emergence of new activities, new goods and services, new jobs, new revenue, etc. For around 40 years, the concept of the innovative milieu and the proximity-based innovation approach have not only increasingly shown, and increasingly well, their relevance as a form of modeling decentralized economic growth, but have also enhanced the technological skills of businesses, including international ones. The geography of innovation, technological districts, the innovative milieu, innovation clusters, technology hubs, etc.: the terms are numerous, and often correspond to different visions of the innovation process and its spatial inclusion[1]. However, beyond their differences, these approaches display one

---

Chapter written by Corinne TANGUY and Dimitri UZUNIDIS.

1 For a presentation of the different types of territorial innovation systems, see [TOR 14].

common point: they believe that innovation can have a strong spatial dimension and that companies can draw an advantage from being located in the same space [AUD 96]. Geographic proximity is therefore the main catalyst for bringing companies together at the regional or local level. These economists do not subscribe to a purely linear innovation framework (which corresponds to the idea that increasing input – R&D expenditure in this case – leads to a growth in output – innovations here), but rather to an interactive vision that pays particular attention to the importance of networks of public and private actors at the level – the territory – recognized as relevant to establishing an innovation policy.

Numerous works are currently being developed on the relationship between entrepreneurs and territories, with the underlying theory of more or less easy access to skills, knowledge, sources of funding and infrastructure, depending on the territories. If the innovative milieu therefore contributes to the innovation performance of companies by providing scientific and technical resources that they can organize, beyond systematic relationships, then the role of economic actors must not be overlooked. By this token, David Audretsch *et al.* [AUD 06] understand the entrepreneur to be the "missing link" between public policies intending to promote innovation and the creation of innovative businesses. We therefore find the image of the entrepreneur as the relay, as described by J.-B. Say, between the expert who produces the knowledge and the worker who applies it to the industry. Similarly, the entrepreneur is also perceived as the unintentional creator of networks of externalities promoting the creation and blossoming of new ideas and new markets [MIN 05].

What, then, are the challenges for an entrepreneur when establishing themselves in a given territory? In order to try and answer this question, in the first section, we will present the concept of an innovative milieu, as well as the proximities at play in regional innovation processes. In the second section, we will see that the role of the entrepreneur is crucial to the fulfillment of the opportunities and resources offered by an innovative milieu, particularly thanks to the networks it is able to deploy. We will conclude with a section dedicated to the policies implemented recently in France with the intention of promoting collaboration and the development of innovation projects in territories, before, finally, we conclude.

2.1. The innovative milieu and proximities

Analysis of the regional innovation system or innovative milieu is relatively recent. Schumpeter placed emphasis very early on the appearance of innovations in clusters or in groups of innovative entrepreneurs, but the question of space was never present in his analyses, and [PER 64], in his analysis of growth hubs, focused on the role of large companies in aiding the expansion of their economic environment. We must consider the works of the evolutionary economists of the 1980s [DOS 88] alongside the idea of the national innovation system to emphasize the importance of economic rules, laws and policies depending on spaces, and to truly take into account adherence to innovation processes in the territory.

2.1.1. *The concept of the innovative milieu*

The local economy (or local production system) can be defined as a geographic space formed as a set of systematic connections between businesses, as well as between businesses, the State and territorial authorities; these systematic connections characterize the localized space according to a certain type of activity and final product. Economists attribute to this economy – which has been observed and studied as a hub for productive relationships able, among other things, to contribute to organization of the territory (which takes place at the national level) – qualifying terms showing the dynamic of combined and complementary relationships between businesses of different sizes at the local level: "local production system", "innovative milieu".

Initially, it was GREMI (Groupe de Recherches Européen sur les Milieux Innovateurs – the European Group for Research on Innovative milieus, a team of researchers from the Institute of Economic and Social Research at the University of Neuchâtel, Switzerland), which, in 1985, departed from the theory that it is regional environments that give rise to different forms of innovation (product innovations, process innovations, organizational innovations, social innovations, innovations in training/qualifications, etc.) [MAI 92]. An *innovative milieu* is defined as a set of actors (businesses, institutions, etc.) that are territorialized, that is, located and based in a territory, and whose actions develop in a multilateral manner [AYD 86]. They generate externalities specific to innovation and are characterized by the convergence of apprenticeships in increasingly reliable forms of common management or resources. In this way, territories differentiate themselves according to the specific knowledge and resources that they have developed historically. The explanation for the emergence of a “successful” region comes from the fact that it is the region that is primarily able to manage its own ability to develop new products, new technologies and new organizations. This assumption, based on regional science, can, to some extent, question the traditional economic theories that, contrastingly, have until now advocated a progress and growth whose causes largely find their justification and origins “outside” (these are the so-called “top-down” development models).

Marshall [MAR 19] showed that the territorial efficiency of this mode of organization lies in what is known today as savings in transaction costs. The concentration in the same geographic area of the main players in the same production system (essentially, the producers on one hand, and the users on the other) allows the facilitation of not only transactions, but also reciprocal relationships of knowledge and trust between different partners. Training and the accumulation of skills can thus form what A. Marshall called an “industrial atmosphere” conducive to shaping the local labor market. This phenomenon – the industrial atmosphere – is linked both to the expertise and professional experience of workers, as well as to the location of businesses in a given territory. In Marshall’s idea of an industrial district, the environment combines population density with proven skills, with a set of actors making up the different links in a production system, and, finally, a practical knowledge strongly rooted in experience. The main feature is not to carry out territorial organization of production according to a hierarchical principle, as in a business, but rather according to a principle of collaboration and cooperation between different production units. In this way, the concept of solidarity between economic actors is very important. The local production system is particularly characterized by the proximity of production units (individual businesses, services, research and training centers, etc.). These units have relationships of varying intensity with each other, which can take very different forms: formal or informal relationships, trading or non-trading relationships, etc. These alliances primarily concern the flow of material, services, labor, technologies and knowledge.

The central theoretical hypothesis of the innovative milieu concept, that is, the socioeconomic environment of a territory shaped by history (“path dependence”), is that it is the product of interactions between companies, institutions and labor. These interactions are by necessity the result of reciprocal synergies (networks, networking, partnerships, etc.) between different local agents (public or private) in economic and industrial development [GAR 93]. Here we can refer, for instance, to the forms of cooperation between companies and research laboratories. Ultimately, it is primarily the immediate socioeconomic, industrial and scientific environment that is involved in the creation of new activities (particularly through entrepreneurship and spinning off) and of innovations (see also [KON 00]). This can only "naturally" take place if certain conditions are respected. These include: the existence, locally, of a group of actors (businesses, research and training centers, public administrations, professional authorities, etc.); the geographically close existence of material, human, financial, technological and computing resources; the existence of specific knowledge that makes a quality production activity possible; the existence of social capital conducive to the creation of local, national or international networks [KRU 93]; and, lastly, the existence of standards, rules and values that favorably determine the conduct of economic actors.

In this way, two dimensions in particular appear to constitute the innovative milieu: technological innovation, learning, experience-based knowledge and resources historically developed as a principle of competitivity; and networks as an organizational principle. The innovation networks in the works of GREMI rely primarily on relationships based on geographic proximity. The innovative milieu establishes the suitable context for its creation and development, as its agents share similar performance, quickly identify partners in their area, exchange information and build trust. The environment participates in building innovation networks and has an impact on their dynamics. Reciprocally, while innovation networks are a-spatial by definition, they enrich the environment and contribute to expanding its creative capabilities [MAI 93].

2.1.2. *Proximity: from space to organization*

Nowadays, the concept of proximity is, for the most part, used as much in industrial economics as in the economics of innovation (see, among others, [RAL 02, BOU 08, MAR 10]). However, the polysemy of the term, just like the variety and extent of its applications ("economies of proximity"), requires it to be used reasonably. Proximity considered *a priori* is shown to be linked to the existence of localized externalities that produce spatial agglomeration effects and territorial dynamics. Other meanings (collected by

Sierra [SIE 97]) were quickly added to this initial meaning of physical proximity, and related to where the field in which the proximity operates (space, organization or institution) is mingled with the content of the proximity relationship (information, training, knowledge or know-how, technology, etc.). Without starting a critical discussion that would lead us too far from the matter at hand, it seems that, for the sake of consistency, we must prioritize the field where the proximity relationship takes place (before giving it content). From this perspective, two kinds of proximity are generally considered when the question of localization is combined with that of organization: spatial proximity, or rather spatio-temporal proximity, and organizational proximity. In this way, beyond spatial proximity, the shared occupation, and the common codes and standards, *the organizational proximity*, play an essential role in the networking of businesses, research centers and public institutions [TOR 05].

The specific features of activities involved in producing new knowledge, and the interactions associated with it, have led economists to introduce, beyond spatio-temporal proximity and organizational proximity, the concept of cognitive proximity [NOO 02, BOS 04]. This returns to a more or less formalized sharing of experiences, presentations, codes, languages and models that results in, and also facilitates, the communication of all information within organizations. Cognitive proximity, by its very nature, occupies a particular place in research activities, but it is also present, albeit in the background, in all sorts of flows of internal or external information to a company. In the case of interactions related to industrial research, cognitive proximity affects not only internal interactions in research centers with other company services, but also interactions with the environment (other laboratories and partners in research and innovation). It is also at this level that the entrepreneur plays a fundamental role as the creator of externalities. For Minniti [MIN 05], these networks of externalities are at the root of creating mental models that enable the ambiguity of information delivered by social environments to be limited. The greater the number of entrepreneurs, the easier it becomes for different information to intersect, and the ambiguity of available information lessens. *In this way, entrepreneurial behavior becomes standard behavior*.

It therefore seems possible that we can propose a three-dimensional approach to proximity, before demonstrating the importance of spatial proximity in generating new activities and launching innovations. The following table shows the main parameters typical of different proximity categories, as well as the area of action and the type of aim related to them. It should be noted at this point that interactions are most often multidimensional. They are a combination of different dimensions from which a main dimension becomes clear; in our case, the dimensions of space and time. *Spatial proximity* is characterized by the reduction of distances (and times) physically separating economic actors; such a condition is conducive to the development of relationships based on acknowledgement and mutual recognition. *Organizational proximity* is defined by belonging to the same organization (business, R&D laboratory, university, service within a given business or administration, etc.), the same network (intra-organizational and/or inter-organizational) or, more broadly, the same "common destiny". *Cognitive proximity* refers to the adherence of various actors to the same concept of innovation, the same paradigm (technological and/or organizational), the same routines, the same heuristics, the same ways of thinking, the same conventions, the same traditions, the same beliefs, the same internal codes, the same languages and/or the same methods of learning, discussing, decision-making and governing. It is therefore found at the very heart of organizations, networks and communities.

Proximity	**Parameters**	**Field of action and aim**
Spatio-temporal	Distance *versus* speed	Movement: flow
Organizational ***intra- and inter-organizational***	Hierarchy *versus* market Intra-company *versus* inter-company Vertical *versus* horizontal Order *versus* contract	Coordination: Strategies, actions, routines
Cognitive	Code *versus* content Context *versus* comprehension (awareness + interpretation) Normalcy of entrepreneurial behavior	Communication Codes, concepts, ideas, knowledge

Table 2.1. *The three categories of proximity*

However, while the idea that production systems benefit from a spatial concentration of research and innovation activities is currently largely accepted, and while geographic proximity seems to be an important component of technological success, we know that, thanks to ever increasing mobility, companies nowadays mainly cooperate with organizations located in different regions. The more open and connected an innovative milieu is, the more it is able to increase its pool of production resources and therefore enhance its attractiveness and capability to produce entrepreneurs. As a consequence, it is believed that a milieu is innovative when it is able to open itself up to the outside world and to take in specific information and resources which its local production system needs for innovation [UZU 10]. Therefore, nowadays, rather than the presence of an innovative milieu, it is the city where services (banking and finance, accounting, insurance, management and legal counsel, advertising) and high levels of infrastructure and equipment, essential to innovation activities and innovative entrepreneurs, are found [HUR 09]. Lastly, proximity is a prerequisite for establishing networks, but it is not always synonymous with the best performances of its actors and can even sometimes hinder them: this is the "proximity paradox". The branch of evolutionist or evolutionary economic geography (for a complete review of this analysis, see the *Handbook of Evolutionary Economic Geography* edited by Ron Martin and Ron Boschma [MAR 10]) supposes that the different types of proximity can lead to path dependence on the part of the innovation networks and, in certain cases, to a regional lock-in (the terms "*path dependence*" and "*lock-in*" are widely borrowed from Darwinism and the evolutionary theory of innovation). According to Bathelt and Schuldt [BAT 08], the milieu's internal links are thus shown to be unfavorable to the transfer of knowledge or innovations when local connections rely only on weak links (or *local buzz*), which do not encourage the transfer of knowledge and innovations. Exchanges with the outside world (*global pipeline*) can therefore prove to be more beneficial in terms of innovation policies, as bringers of radical innovations are able to communicate with distant companies.

2.2. The innovative milieu and entrepreneurs

The creation of innovative or more traditional businesses is determined, as we have just seen, by the wealth of the "milieu". However, a business is nothing without an entrepreneur, namely the economic agent who takes the decision to create a business. It is the entrepreneur who makes use of all of the resources in its territory and contributes to the creation of new technologies and jobs.

2.2.1. *The innovative entrepreneur: networks and resources*

Feldman [FEL 01] lists the factors that are traditionally associated with entrepreneurship in the literature: the availability of financial resources, the support of social and relational capital, the existence of support services and entrepreneurial expertise, and, lastly, the presence of universities and research centers. Within the framework of the network economy, the existence of a space created by proximity relationships is beneficial to innovative entrepreneurship: transport infrastructure, communications infrastructure, telecommunications, education, engineering, etc.; contractual, informational, cognitive, trust-based and cooperation-based relationships; a dense business network; a mechanism for fiscal and financial aid (based on the ease or difficulty of accessing public aid, the development of venture capital, etc.), and so on [UZU 10]. Subsequently, all milieus either are or are not becoming innovative. Some are even downright suppressive, and it is essential for a territory to ensure the maintenance and renewal of its key skills [MEN 06], that is, the skills critical for its survival. This issue is of even greater importance when these territories include former industrial sectors, but must also enter into international competition. However, a lack of financial resources and/or public support [DEP 10], or even the existence of conflicts or the difficulty of building and sharing a common project between different actors [TOR 12], can consequently act as obstacles to entrepreneurial dynamics. The ability to set about taking action is therefore the product of the variety and wealth of resource potential that is mobilized or able to be mobilized by the entrepreneur [UZU 10]. The composition of this resource potential depends on external factors relating to both the business and the entrepreneur (namely, the wealth of the innovative milieu). It is the entrepreneurs themselves who, by being adaptable and seizing new opportunities, create the factors and conditions that favor their businesses and contribute to the development of these external resources [FEL 01].

Entrepreneurs, against a background of innovation, particularly use their social networks to confirm the existence of a business opportunity, that is, to prove that there is a possibility of financial gain [CHA 10]. Beyond the initial financial resources that allow a project to begin, entrepreneurs are in fact confronted by numerous difficulties, which are overcome thanks to the mobilization of relational resources. Concerning researching information or acquiring resources, entrepreneurs rely largely on strong relationships, as these appear to be both more easily accessible and trusted [CHA 10].

Emphasis is therefore placed on the human capital of entrepreneurs. Thanks to their network of social relationships, they are able to access resources (in knowledge, finance, etc.) and information that they do not have. The entrepreneur is embedded in a network of social relationships [GRA 06]. The formation of business networks is also a means of reducing the risks related to market uncertainty. The more protective the networks are, the easier it is for the entrepreneur to act [UZU 10]. However, while it is always preferable to rely on social relationships to launch an activity, over time these can become a problem for the activity's development if it translates into overly large obligations for the business (interference in the management of the company, an obligation to hire a family member or give "discounts" to relatives, etc.) [GRA 03]. Social networks affect economic activity for three main reasons: (1) they influence the flow and quality of information, (2) they are an important source of rewards and sanctions that have a greater impact when they come from people we know personally and (3) they create trust, which is defined as the belief that the other economic actors will do "the right thing", despite a clear incentive to act otherwise. Economic action is well integrated into systems of personal relationships and individual networks always have a local aspect, although the importance and nature of local links in a given place vary according to the movements of the actors and their characteristics (frequency, periodic or not, etc.). In this way, localized innovation networks can be explained less by the operational need to exchange information in person, than by the cooperation established between researchers or engineers who attended the same university or belong to the same social or familial network [GRO 03]. In other words, the effects of proximity can vary depending on context, and, if the individuals were fully mobile, the effects of proximity would disappear.

The entrepreneur is a figure, a concept and a function difficult to pin down with existing theoretical tools [TOU 06]. The personal qualities and the personality of the entrepreneur undoubtedly play an important role in the decision to create or take on a small business. However, the function and act of entrepreneurship are defined by the macro-systemic dynamic of accumulation and profit. On the other hand, the creative act is part of a social framework, and the fact remains that "*one is not born an entrepreneur, but becomes one*"; a person becomes an entrepreneur by mobilizing his or her resource potential, formed of capital, knowledge and relationships [BOU 01, BOU 06]. The capital is for investment and operations; the knowledge is for making choices and decisions; and the relationships are for finance, collaboration and carrying out production.

2.2.2. *Policies for creating innovative milieus and the emergence of an innovative entrepreneurship*

A company's territorial anchorage – its links to the local community – allows it to build up a store of resources (and sometimes a market) to reduce the costs of its investments in a constantly changing economy. However, this territorial anchorage is dependent on the quality of the aforementioned store in relation to the company's expectations in terms of innovation and commercial expansion. This is the why there is a need for governments and regional authorities to create a system for organizing the development of resources with the aim of generating multiple innovation processes, taking account of the competition and cooperation that exists even between actors in the open economy (see in particular [AMA 97, PIT 05]). It is a question of a system offering production resources able to produce technology-based entrepreneurship and attract large companies with proven track records in innovation [MAD 05, UZU 04].

For two centuries, public authorities and large companies have implemented an important administrative mechanism for creating businesses (aimed at increasing the "resource potential" of the entrepreneur) in order to fill the gaps caused by economic activity and to adopt means of trying out new market values and new social practices. Therefore, in the year 2000, the Organisation for Economic Cooperation and Development (OECD) encouraged governments and regional institutions to intervene in the economy to "foster entrepreneurship" [OCD 99]. Since the 1980s, various efforts in regional innovation systems (innovative environments, localized production systems, clusters) have given rise to the implementation of public policies [HER 03]. The mechanism for local production systems, and logistical and financial support for selected networks of companies was launched in France by DATAR (the Interministerial Delegation of Land Planning and Regional Attractiveness) in 1998. In 2009, DATAR launched a policy of business clusters concerning groups of companies with shared projects at the regional level, particularly in sectors with low R&D activity. The most striking example of implementing a structural policy to aid entrepreneurship and the economic appeal of "milieus" is the policy of creating competitiveness clusters [FON 05], which is not only found widely throughout the industrial world (for Europe, see [PEY 02, UZU 06]), but also found in some large developing countries (India, China, etc.). The policy of competitiveness clusters which has been implemented in France since 2004 has aimed to encourage collaborative research and development projects between local actors in the same territory.

According to the official definition[2], a competitiveness cluster "*brings together companies, educational establishments and public and private research institutions in a given area in a partnership-based approach intended to build synergies around collaborative innovation projects. This partnership is organized around a market and its related technological and scientific field and must research the critical mass to achieve not only competitivity, but also international visibility*".

Competitiveness clusters combine attractiveness (agglomeration effects) and diffusion (dispersion effects), two key factors for the creation and dissemination of innovation. In this way, a territory is conducive to entrepreneurship, and therefore, to supplying the "network economy", when it is built economically by significant potential for innovation, a varied and rich financial system, a large range of services in industry, a system for protecting intellectual property and confidentiality, a dense network of SMBs, access to a large regional/national market, and incentivizing policies in research and innovation.

Small innovative businesses break into networks made up of large groups, often coordinated in relation to their territorial presence [UZU 03]. This is the case of science parks where companies with wide-ranging technological and commercial advantages create small, specialized companies (kinds of research laboratories) by themselves, which are managed by entrepreneur-researchers. In other cases, it is the frameworks of a large business which are asked by the parent company to create a business experimenting in new technologies (intrapreneurship). It is no longer unusual to witness the appearance of "speculative entrepreneurship" [TID 05] for the benefit of large companies. It involves highly qualified individuals who, with the aid of subsidized loans and state handouts, create a business in order to resell it to a larger one, once the product or production process has been developed. The connection of small specialized business to groups is carried out by means of a financial and compliance-based strategy adapted to the subject. Venture capital companies (companies acquiring stakes in the capital of a newly created business), *business angels* (wealthy individuals who invest in innovative projects) and other financiers (pension

2 Government website: http://www.competitivite.gouv.fr.

funds are very active here) commit the capital of savers to new companies in information technology and biotechnology.

2.3. Conclusion

It is the local as well as the entrepreneurial aspects of innovation that reveal the mechanics of creating and appropriating all the ingredients (capital) in productive activity. The issue of excludability has become crucial to the operation (localization or creation) of a business. A business tends to make use of its environment before investing, for example, in all stages of technology creation; this can be explained by the fact that investments in acquiring (appropriating) production resources are less expensive than investments dedicated to creating those resources. This also explains the attractiveness (in an open economy) of a region with plentiful scientific and technological resources. The creation of innovative or more traditional businesses is determined by the wealth of the "milieu". If the factors related to, *inter alia*, education, the environment, medical care, finance, infrastructure and housing affect the trajectory of the marginal cost of a business or an activity, then they also influence the return on capital invested. The ideas of the "network" and the "innovative milieu" have therefore emerged to make themselves essential to economic analysis and observation.

The concept of the innovative milieu highlights a strong principle at the level of innovation systems; it reinforces the idea that the innovation capabilities of businesses are intimately linked to the social, economic and political factors around them [CRE 00]. The "innovative milieu" therefore most often refers to the capability of a local economy to produce innovations through the emergence of new businesses and the localization of existing businesses in its geographic area, where the industrial emphasis on research organizes the creation of small, innovative companies. In this vein, the local economy appears to be a regional system for developing all kinds of capital and commercial exchanges. As it enjoys some autonomy in the (legal and economic) organization of production resources, its major feature is the creation and development of specific resources and the achievement of particular combinations of these specific resources (see the case of Sophia-Antipolis studied by Elidrissi and Hauch [ELI 08]). These are made up of capital and work with specific content and forms in relation to given sectors and activities: they are not only technologically, financially and demographically specific, but also specific in terms of skills, qualifications, level of education, etc. The local economy therefore becomes an "innovative milieu" that reduces the risks related to the uncertainty of any given investment and begins the innovation process, including by means of creating business and accommodating existing technology companies.

However, the business is nothing without the entrepreneur, namely the economic agent who takes the decision to create a business. It is the entrepreneur who draws on the resources of his or her region and contributes to creating new technologies and jobs. Innovation, as suggested by Audrestch, Keilbach and Lehmann [AUD 06] is an endless process, as new business founders create a new activity by valuing ideas that have been ignored by other entrepreneurs. The investment of millions of euros or dollars to finance R&D is fruitless in the absence of entrepreneurs. Moreover, the growth of individual initiatives contributes to turning the entrepreneur into a behavioral model that other economic agents can come to imitate, simultaneously contributing to the creation of new innovations. Within the defined framework of a particular territory, economic agents have created a network of social relationships [GRA 06] of all kinds, thanks to which they have access to the resources they need to complete their project, including information resources.

2.4. Bibliography

[AMA 97] AMABLE B., BARRÉ R., BOYER R., *Les systèmes nationaux d'innovation à l'ère de la globalisation*, Economica, Paris, 1997.

[AUD 96] AUDRETSCH D.B., FELDMAN P.M., "R&D spillovers and the geography of innovation and production", *American Economic Review*, vol. 86, pp. 630–640, 1996.

[AUD 06] AUDRESTCH D.B., KEILBACH M.C., LEHMANN E.E., *Entrepreneurship and Economic Growth*, Oxford University Press, Oxford, 2006.

[AYD 86] AYDALOT P., *Les milieux innovateurs en Europe*, Economica/GREMI, Paris, 1986.

[BAT 08] BATHELT H., SCHULDT N., "Between luminaires and meat grinders: international trade fairs as temporary clusters", *Regional Studies*, vol. 42, pp. 853–868, 2008.

[BOS 04] BOSCHMA R., "Proximité et innovation", *Économie Rurale*, vol. 280, pp. 8–24, 2004.

[BOU 01] BOUTILLIER S., UZUNIDIS D., "L'utilité marginale de l'entrepreneur", *Innovations*, vol. 13, no. 1, pp. 17–42, 2001.

[BOU 06] BOUTILLIER S., UZUNIDIS D., *L'aventure des entrepreneurs*, Studyrama, Paris, 2006.

[BOU 08] BOUBA-OLGA O., CARRINCAZEAUX C., CORIS C., "La proximité, 15 ans déjà ! Première partie: propositions théoriques", *Revue d'Economie Régionale et Urbaine*, vol. 3, pp. 279–287, 2008.

[CHA 10] CHABAUD D., NGIJOL J., "Quels réseaux sociaux dans la formation de l'opportunité d'affaires?", *Revue française de gestion*, vol. 206, pp. 129–147, 2010.

[CRE 00] CREVOISIER O., "L'approche par les milieux innovateurs: état des lieux et perspectives", *Revue d'Économie Régionale et Urbaine*, vol. 1, pp. 153–166, 2000.

[DEP 10] DEPRET M.H., HAMDOUCH A., MONINO J.L. *et al.*, "Politiques d'innovation, espace régional et dynamique des territoires: un essai de caractérisation dans le contexte français", *Innovations*, vol. 33, no. 3, pp. 85–104, 2010.

[DOS 88] DOSI G., FREEMAN C., NELSON R.R. *et al.* (eds), *Technical Change and Economic Theory*, Pinter, London, 1988.

[ELI 08] ELIDRISSI D., HAUCH V., "Les systèmes régionaux d'innovation: superpositions et articulations. Le cas de Sophia Antipolis", *La revue des sciences de gestion*, vol. 233, pp. 59–66, 2008.

[FEL 01] FELDMAN M.P., "The entrepreneurial event revisited: firm formation in a regional context", *Industrial and Corporate Change*, vol. 10, no. 4, pp. 861–891, 2001.

[FON 05] FONTAGNÉN L., LORENZI J.H., *Désindustrialisation, délocalisations, Conseil d'analyse économique*, La Documentation française, Paris, 2005.

[GAR 93] GAROFOLI G., "Economic development, organization of production and territory", *Revue d'Économie Industrielle*, vol. 64, pp. 22–37, 1993.

[GRA 03] GRANOVETTER M., "La sociologie économique des entreprises et des entrepreneurs", *Terrains et travaux*, ENS Cachan, vol. 4, pp. 167–206, 2003.

[GRA 06] GRANOVETTER M., "L'influence de la structure sociale sur les activités économiques", *Sociologies Pratiques*, vol. 13, pp. 9–36, 2006.

[GRO 03] GROSSETTI M., BÈS M.-P., "Proximité spatiale et relations science – industrie: savoirs tacites ou encastrement (Polanyi ou Polanyi)? Réflexions à partir d'une étude sur les collaborations CNRS – entreprises en France", *Revue d'Economie Régionale et Urbaine*, vol. 5, no. 2, pp. 777–788, 2003.

[HER 03] HÉRAUD J.A., "Régions et innovation", in MUSTAR P., PENAN H. (eds), *Encyclopédie de l'innovation*, Economica, Paris, 2003.

[HUR 09] HURIOT J.M, BOURDEAU-LEPAGE L., *Economie des villes contemporaines*, Economica, Paris, 2009.

[KON 00] KONSTADAKOPOULOS D., "Milieux innovateurs et apprentissage dans le Sud-Ouest de l'Angleterre", *Innovations*, vol. 11, pp. 139–154, 2000.

[KRU 93] KRUGMAN P., "First nature, second nature and metropolitan location", *Journal of Regional Science*, vol. 33, pp. 129–144, 1993.

[MAD 05] MADEUF B., CARRÉ G., LEFEBVRE G. *et al.*, "TIC et économie de la proximité: organisation et localisation de la R&D au sein des entreprises globales", *Innovations*, vol. 21, pp. 243–272, 2005.

[MAI 92] MAILLAT D., PERRIN J.C. (eds), *Entreprises innovatrices et développement territorial*, GREMI/EDES, Neuchâtel, 1992.

[MAI 93] MAILLAT D., QUEVIT M., SENN L. (eds), *Réseaux d'innovation et milieux innovateurs: un pari pour le développement régional*, GREMI/EDES, Neuchâtel, 1993.

[MAR 19] MARSHALL A., *Industry and Trade*, Macmillan, London, 1919.

[MAR 10] MARTIN R., BOSCHMA R. (eds), *The Handbook of Evolutionary Economic Geography*, Edward Elgar, Cheltenham, 2010.

[MEN 06] MENDEZ A., MERCIER D., "Compétences-clés de territoires. Le rôle des relations interorganisationnelles", *Revue française de gestion*, vol. 164, no. 5, pp. 253–275, 2006.

[MIN 05] MINNITI M., "Entrepreneurship and network externalities", *Journal of Behavior and Organization*, vol. 57, pp. 1–27, 2005.

[NOO 02] NOOTEBOOM B., A cognitive theory of the firm, Paper for ESNIE Workshop Alternative Theories of the Firm, Paris, 2002.

[OCD 99] OCDE (Organisation de la coopération et du développement économiques), Stimuler l'esprit d'entreprise, OECD, Paris, 1999.

[PER 64] PERROUX F., *Industrie et création collective, Tome 1: Saint Simonisme au XXème siècle et création collective*, Presses Universitaires de France, Paris, 1964.

[PEY 02] PEYRONI J., *Le schéma de développement de l'espace communautaire*, La Documentation française, Paris, 2002.

[PIT 05] PITELIS C., SUGDEN R., WILSON J.R., *Clusters and Globalisation*, Edward Elgar, Cheltenham, 2005.

[RAL 02] RALLET A., TORRE A., "Proximité Géographique ou Proximité Organisationnelle? Une analyse spatiale des coopérations technologiques dans les réseaux localisés d'innovation", *Economie Appliquée*, vol. 1, pp. 147–171, 2002.

[SIE 97] SIERRA C., "Proximité(s), interactions technologiques et territoriales", *Revue d'économie industrielle*, vol. 82, pp. 7–38, 1997.

[TID 05] TIDD J., BESSANT J., PAVITT K., *Managing Innovation*, Wiley, Chichester, 2005.

[TOR 05] TORRE A., RALLET A., "Proximity and localization", *Regional Studies*, vol. 39, no. 1, pp. 47–59, 2005.

[TOR 12] TORRE A., BEURET J.E., *Proximités Territoriales*, Economica, Paris, 2012.

[TOR 14] TORRE A., TANGUY C., "Les systèmes territoriaux d'innovation: où en est-on?", in BOUTILLIER S., FOREST J., GALLAUD D. (eds), *Principes d'Economie de l'Innovation*, Peter Lang, Brussels, 2014.

[TOU 06] TOUNES A., FAYOLLE A., "L'odyssée d'un concept et les multiples figures de l'entrepreneur", *La revue des sciences de gestion*, nos 220–221, pp. 17–30, 2006.

[UZU 03] UZUNIDIS D., BOUTILLIER S., "L'Etat, les entrepreneurs et la mondialisation", in VAN CROMPHAUT M. (ed.), *L'Etat à l'ère de la mondialisation*, Economie et Innovation, L'Harmattan, Paris, 2003.

[UZU 04] UZUNIDIS D. (ed.), *L'innovation et l'économie contemporaine*, De Boeck, Brussels, 2004.

[UZU 06] UZUNIDIS D., LAPERCHE B., La territorialisation des aides et de la politique européenne de R&D. Quelles perspectives pour la France?, ESPON, European Union, 2006.

[UZU 10] UZUNIDIS D., "Milieu innovateur, relations de proximité et entrepreneuriat. Analyse d'une alchimie féconde", *Revue canadienne de science régionale*, vol. 33, pp. 91–106, 2010.

3

Start-up Founders and Support for Technology Entrepreneurs

The support needs of start-up founders are numerous. A start-up is an innovative company with strong ties to new technology and strong market prospects. The entrepreneur often concentrates on their core business by making use of their experience, knowledge and skills. They find it difficult to know how to deal with stakeholders with multiple and varied expectations. The support needs are significant, particularly as complementary skills are necessary to create and launch a complex business. In this sense, the support can include five levels of intervention [CAS 96, EAS 14, RIE 11]: expertise (posing solutions), coaching (helping to find solutions), teaching (transfer of knowledge and know-how), financing (helping to identify and optimize the sources and kinds of financing) and entering the market (supporting the successful realization of the project).

As for the spotter (the person providing the support), they target three main factors crucial to the success of a start-up launch. These concern the founder, the project and the market prospects. The spotter's job consists of addressing particular problems, transferring knowledge or specific skills, and encouraging or motivating the entrepreneur with a view to improve personal and organizational performance. The logic underpinning the founder's actions is strongly affected by their needs to *be*, to *know* and to *be able*, to the extent that it can reconstruct the essence of their entrepreneurial identity. These three needs are the basic components of the founder's identity which

---

Chapter written by Gérard A. KOKOU DOKOU.

the spotter must work into their approach based on expert advice [CAR 09, HOA 10, MUR 07]. These same components of entrepreneurial identity underpin the power of the founder's actions in terms of theories around resources and their application [SAR 08, TEE 07].

The aim of this chapter is to pinpoint the content of the three components of the entrepreneurial identity of technology creators. Our analysis focuses on start-up entrepreneurs. It is primarily a question of revealing the interactive processes that connect the technology creator and the organization he or she founds. These interactive processes draw on the overlapping nature of various phenomena, with the most important of these being the motivational and skills-based elements underpinning the initial project, socialization in a constantly changing technological context, identification of an organization hungry for resources and high-quality skills, and sharing relationships with very demanding financial and technological communities. Deciphering such processes helps to enrich support teaching lead by advisers who are experts in technology entrepreneurship.

In order to achieve this research objective, a qualitative study has been carried out in the form of a focus group bringing together eight start-up founders. Conversations were recorded and transcribed in their entirety in March 2015. In this way, the body has been carefully examined using methods of semantic and structural analysis [GHI 78, GHI 85, MUC 74]. The interpretative approach leads to emphasis being placed on the importance of the components of identity at the level of support for start-up founders.

Before this crucial result, the first part of our work will show a rapid enlightenment with a theoretical scope. This has energized recent development in managerial research which concentrates on the use of key resources, value systems and knowledge as a way towards future understanding of the components of the entrepreneurial identity of a business founder. The second part concerns an inductive approach. This is related to the identity components and operational logic of the technology entrepreneurs surveyed. The highlighted identity components act as levers for action for the technology creator's expert adviser.

3.1. Putting entrepreneurial identity into theoretical perspective

3.1.1. *The needs for existence, knowledge and ability at the heart of entrepreneurial identity*

Self-identity is the product of various personal expectations. These concern digressions and recurrences, repression and enablement of repression to form a need to be. Its construction occurs in a dynamic, multidimensional and complex manner. It is spread through successive periods of socialization: personal life, schooling, professional experience, collective values and perceptions, etc. These socialization processes are both determining factors in constructing identity and influences on that identity [VIB 11].

At the same time, the progressive construction of identity contributes to guiding an individual in their environment, giving meaning to their experiences and providing indications for action. If the individual chooses an entrepreneurial direction, they imagine situations and activities in this area and adjust their behavior accordingly. This allows them to move away from a situation where they conform to the expectations of others (because of their roles in families, friendships or workplaces). Eventually, they create a new self-image that takes the form of a professional identity.

This configuration incorporates various elements and gradually develops through successive analyses and self-reorganization. For Erikson [ERI 68], identity is therefore the product (unified, consistent, stable) of a continually open and interactive process of self-creation, which must be validated by the subject and recognized by others. It is introduced in dynamic terms by the competition between a demand for harmony and an acceptance of discord [RIC 90]. Identity constantly readjusts itself, and it is through engagement at the heart of actions and in relationships with others within interactions that the subject develops and constructs their identity. It is an exploratory process involving trial and error. Unusual situations that may be encountered can provoke the creation of new behaviors, but those must be justified, validated and ratified.

In constructing their life story, the subject reconciles the diversity and variability of events in a uniform interpretation, the result of negotiation with oneself, but whose terms of agreement are ordered and fixed differently for every individual [STR 92]. Discernible in everyday actions and occurrences, the subject's identity can therefore have many aspects and take many forms [RUA 04], and it can also be the object of reassessment, reinterpretation, reformulation or even destruction. The relevance of multiple representations for the purposes of assimilation/differentiation is real [DEL 99, MAR 97, TAP 97]. It is a question of representations that are self-gratifying and open to learning.

Learning validates the need for knowledge. Indeed, the motivation for constructing a professional identity truly occurs when there is a structuring dynamic using concrete elements that are ubiquitous within the support framework: new representations, identifications gained through new experiences, new skills and the use of all of these acquisitions through the support process. All situations that are experienced by the individual strengthen, consciously or unconsciously, the identity-creation process. The individual is subject to endogenous learning factors (their personality, value system, standards, etc.) and exogenous ones (a business, university, relationships with various colleagues in the business), which are able to push the individual towards developing their knowledge, behavior, points of reference, projects and management practices. The need to be able to do something therefore finds a suitable solution. In other words, the driving elements do not work separately in the identity process but rather in an interactive manner, setting in motion an interplay of expressions, relationships, developments and regulations unique to the emergence of a system favorable to successful professional actions. Entrepreneurship fits into this system with specific skills.

The construction of an entrepreneurial identity therefore requires a system of interactions between:

– personal motivation and the tasks assigned to the organization created;

– the necessary entrepreneurial knowledge and skills;

– the elements facilitating the entrepreneurial process and the founder's aspirations.

These interactions refer to three kinds of entrepreneurial motivation through the prism of the three needs mentioned initially: a need for fulfillment expressed in the professional domain, a need for belonging in the affective domain, and a need for ability, meaning ability in the sense of knowledge acquisition, which is part of the cognitive domain. These same interactions allow us to understand and explain how the environment (school, support structures, etc.) affects the self (or the individual) and how the self affects the acquisition and construction of knowledge and know-how in creating entrepreneurial behaviors [OBR 16]. Self-knowledge, motivation and entrepreneurial skills develop as a result of support.

In other words, all individuals continually carry out a process of self-reevaluation which can occur at three moments [BRO 94]: in cases of classification (positioning, roles in relation to other people: head of business/employee); narrative (business-based or personal relationships between the individual and others: domination or cooperation) or attribution (self-evaluation in terms of taking responsibility compared to others). The discussion of identity therefore also makes a connection between the level of the individual (egocentric reasoning) and of the group (individual reasoning and the practices of presenting oneself within a group), including in the creation of a new business [HOG 00].

Ultimately, identity construction reflects personal history and is one of the key forms of individual practices and representations. Personal history includes various different elements: the interaction of the individual with their parents, the learning of roles related to their gender, the education received in their environment, etc. The act of beginning to do something provides the individual with the freedom to pursue their own motivations, dreams and desires when they create a new business. This entrepreneurial activity is infused with purpose, as it is an expression of individual identity or a notion of self that serves to mobilize motivations, skills and resources [CAR 09, HOA 10, KIM 79, MUR 07].

3.1.2. *Expanding towards a resource-based approach*

The resource-based approach has expanded to the field of entrepreneurship [ALV 01, ALV 02, ART 06, ZHA 06]. We can identify specific personal abilities in the entrepreneur that form a real and critical resource. It is a matter of "meta-skills" with regard to the dynamic version of

Teece's *resource-based theory* [TEE 07]. These meta-skills determine the entrepreneur's ability to act in changing environments. They take the measure of an open and dynamic entrepreneurial identity in Sarasvathy's model of "effective entrepreneurship" [SAR 08]. In effect, they are present in:

– the consubstantial capabilities of the individual entrepreneur: the need for existence in their identity (*who am I?*), and their expertise and need for knowledge (*what do I know?*);

– spotters and other stakeholders (*whom do I know?*) with interdependence between the entrepreneur and the players involved, and interdependence that determines their ability to act or the need to be able to act (*what can I do?*).

During the set-up process carried out by the entrepreneur (who makes choices in accordance with potential impacts of certain methods, and not according to a predetermined objective), stakeholders play a central role as it is their commitment on which the possibility of pursuing the resource development process depends. Similarly, the reference to identity leads to the matter of the values of the individual entrepreneur, recognizing that the roles composing a person's identity as well as their values are determined socially and are therefore exposed to a range of societies and their cultures.

An individual's identity is in fact constructed in accordance with two aspects, the relationship with the self and relationships with others, which the psychological theory of structural identity analyzes in terms of roles. This results in an "*individual internalization of roles in line with social expectations which focuses on behaviors linked to a function or status*" [MUR 07]. The identity of an individual entrepreneur will therefore be a construct made up of "*multiple interacting identities*" (spouse, mother or father, leader of an organization, etc.), as well as the entrepreneurial identity. However, it finds unity in the fact that the individual also carries mental structures that exist at a higher abstract level than attitudes. Yet, according to Hitlin [HIT 03], values develop in social contexts, use symbols to refer to significant cultural elements and form an aspect of the self that is both necessary and fundamental, although not coercive. Values, which are calibrated in social settings, are deeply rooted in the individual and act, according to the author mentioned, as a "*cohesive force for social identity*". They are often strongly molded by religion [AUD 07]. In some contexts, there is a direct link between the religious aspect, on the one hand, and the individual's inclination to become an entrepreneur or the local structures for entrepreneurial action, on the other [DAN 10], as well as the generation of the necessary knowledge. This is a means of making the different components of entrepreneurial identity dynamic, as shown in the figure below.

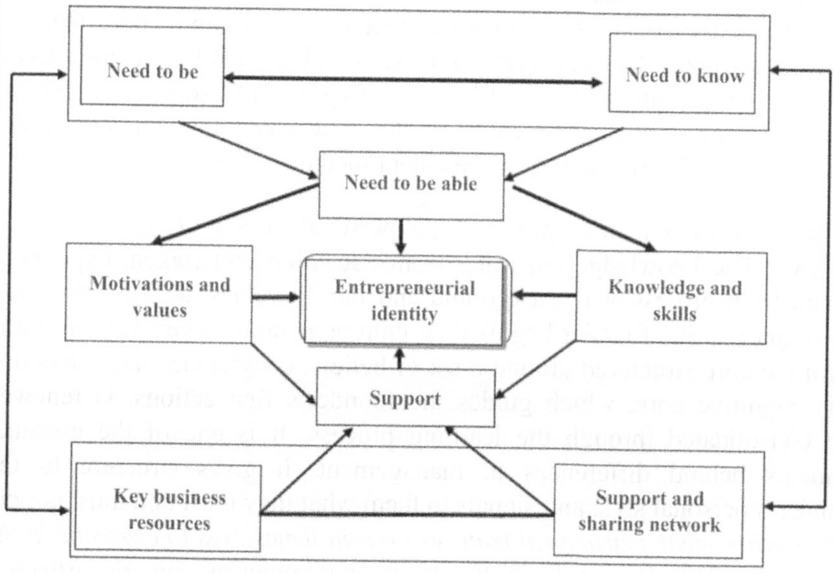

Source: Kokou Dokou G.A. (2014, p. 55) Functional components of entrepreneurial identity

Figure 3.1. *Functional components of entrepreneurial identity*

The content of our figure suggests the main operational levers that make up entrepreneurial identity. The process of building this identity is assisted by entrepreneurial support and its influence on business creation. Entrepreneurial identity is constructed through the subject's affiliation with various groups and could be defined as a polymorphous, dynamic structure whose constituent elements are the psychological and social aspects related to the relational situation of an individual as a social actor at a given moment. In a business creation situation, the relevant elements vanish for the benefit of other elements when the situation alters under the influence of entrepreneurial support. The entrepreneurial support staff, in one way or another, shape the founder's motivations and vision. They give the founder the capability to work on his or her own definition of self.

Through this support process, the individual determines their conscious or unconscious behaviors adapted to be put in place to achieve their "complete identity victories" against an opponent who may be the individual themself, other people in concrete interactions or the social environment. One of the aims of the individual founder is to gain recognition of their existence in the social and economic system. This involves both that this system recognizes their belonging and specific place, and that it subjectively senses this recognition. Assisted by the entrepreneurial support process by affirming their identity, the creator, by the same token, maintains the frame of reference in their behavior, management methods, products, etc.

In the entrepreneurial process, the personal action of the founder is decisive. The knowledge emerging from education undertaken, experience gained from professional background and family values are key. We must not forget that the founder begins their entrepreneurial career with an initial cognitive core structured around a set of beliefs, convictions and behaviors. This cognitive core, which guides the founder's first actions, is renewed and consolidated through the learning process. It is one of the essential elements behind differences in management. It gives structure to the founder's personal style and signals to them what they must and must not do. "*The environment's structural features present themselves to everyone in the same way, but do not have the same consequences for all. Effective entrepreneurs differentiate themselves through their ability to detect and use opportunities for their own ends, while protecting themselves against difficulties*" [BOU 94, p. 117].

Such an ability is the product of multiple skills strengthened by the entrepreneur's continued training. The training activities often have repercussions that go beyond the completion of the projects considered. They contribute to encouraging the development of entrepreneurial culture and strategic training. The best balance can be found between theory and practice. This establishes the global nature of support which is based on:

– the facilitators and strategic risk factors of the entrepreneurial process;

– the key resources to be mobilized for the fledgling business;

– the entrepreneur's networks for support and sharing.

These three dimensions take on a particular connotation in the case of start-ups. The start-up is generally defined as a young business with a strong development potential and which requires significant investment with the aim of generating rapid growth with a high level of risk. It is often found in the field of digital technology or new technologies. It is above all a state of mind, a way of seeing the world that best characterizes the entrepreneur behind a start-up. Regardless of its field of activity, the start-up has the ambition of capturing part of the value of the market and gaining a dominant position in it. In this regard, Blank and Dorf [BLA 13, p. 25] have defined it as "*a temporary organization in search of a viable and profitable business model that enables growth*". Eric Ries, meanwhile, specified that the start-up is "*a human institution designed to create a new service or product in extremely uncertain conditions*" [RIE 11, p. 27]. The three other identity dimensions of such an organization concern the key characteristics of the entrepreneur, the particular components of the business and the environmental context [LIT 98, SOL 00, TÖD 03].

These three dimensions allow the technology businesses that succeed to be well differentiated from those that fail. A technology entrepreneur knows how to combine technical, human and financial resources. The technological resources are crucial. However, in order to make use of them, there must be highly qualified human resources with technical skills and knowledge. The financial resources allow the start-up to have the best associates to be able to develop and make use of technological resources.

3.2. Demonstration of the components of the identity of a start-up entrepreneur

3.2.1. *Creating the body of research and operational tool used*

By "body" we mean a set of collected texts outlining a research objective. This has guided several works of research and empirical studies that we have led[1]. At the level of the latter, the main topics of exchange concern: their motivations and interest in being entrepreneurs, their values and objectives as entrepreneurs, their main skills and those essential to the job of

1 The latest research work was made up of informative material from a focus group of eight start-up founders from the north of France. The focus group method consisted of bringing together 6–12 voluntary representatives for the specific subject of research. The basic principle was respected with our eight participants.

technology entrepreneur, the advantages and disadvantages of being an entrepreneur, the facilitators of their path to entrepreneurship, their support needs, the challenges and aspirations for the next 5 years, and their perception of the situation of entrepreneurship in general and in their region, in particular. All of the discussions, which lasted three hours on average, were recorded and transcribed in their entirety to form the body of research.

Our analyses are based on the ALCESTE system (*Analyse des Lexèmes Co-occurrents dans un Ensemble de Segments de Textes* – Analysis of co-occurring lexemes in a set of text segments). This is a complete methodology for data analysis, which is defined by three approaches: a lexical approach by means of discursive statistics and lexicometry, a content analysis approach consisting of splitting the text into contextual units and a data analysis approach classifying these contextual units [REI 98]. In this regard, ALCESTE becomes part of the general research framework for linguistic data analysis [BEN 82, LEB 00]. It largely consists of studying the laws of vocabulary distribution in a body underpinned by multidimensional or varied approaches, such as the factorial analysis of correspondence and ascending hierarchical classifications (A.H.C.) or descending hierarchical classifications (D.H.C.).

The operating mode of the ALCESTE system is a process consisting of four stages. *Stage 1* is intended to recognize the discourse of each of the bodies investigated using a number of lexical tools. During *stage 2*, ALCESTE splits the body into text units, which are called elementary contextual units. This is followed by analysis of the vocabulary specific to these units and their classification. It is on the basis of the categories identified that perceptions and operating logic of the bodies under investigation are elucidated in order to be interpreted. *Stage 3* provides the analyst with the major files with the most significant results. These contain the different categories established, their mutual dependence and the dominant vocabulary, such as the notable presences and absences of shapes within categories, as well as the elements of identifying those investigated. Lastly, *stage 4* leads to definitive classification. All of this encourages the analysis and interpretation of the set of results.

The four stages of exploiting the entire body have revealed the specific features of the operating logic unique to the start-up entrepreneurs being surveyed. These specific features strongly correlate (chi^2 test)[2] with their perceptions of the impact of new technologies, particularly digital technology. They are made clear in the developments that follow.

3.2.2. *Revealing the main identity components*

The graph below (Figure 3.2) translates one of the key results from the exploitation of the discussion data used. It clearly indicates the importance of technology in the creator's entrepreneurial process.

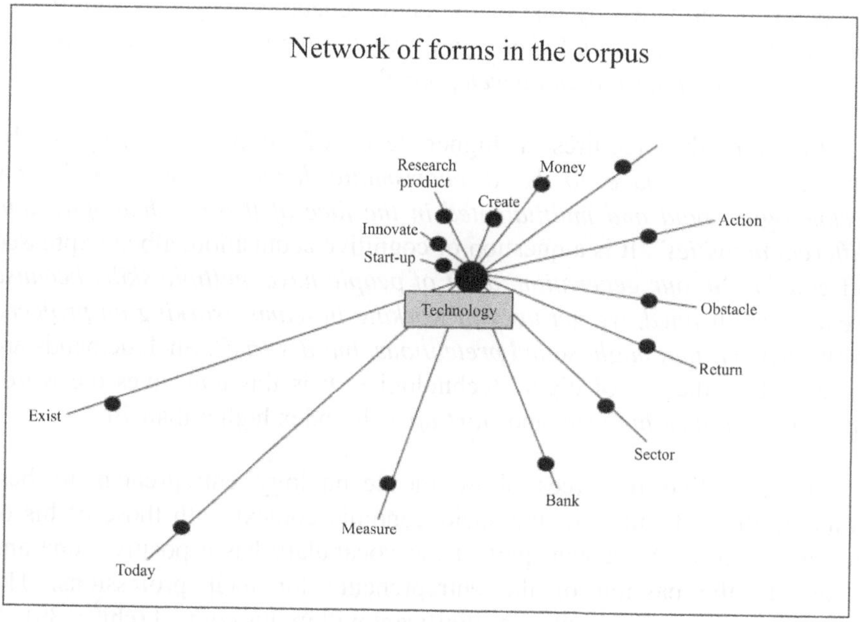

Figure 3.2. *Graph of the major lexical shape of the body studied*

2 To explain, the coefficient chi^2 (metrics of chi^2) enables appreciation of the significance of the link or association between two classified contextual units (or themed units) or two categories. The most important pairings must have a chi^2 value higher than 4.

Technology is at the heart of the entrepreneurial process of the founder being surveyed. It is really a question of start-ups that galvanize human and financial resources, and guidance. Banking partners play an important role. In order to steer the necessary actions towards the success of entrepreneurial ventures, it is crucial to have the advice of specialists or experts. Work in teams and communities is vital. This allows knowledge and skills to be adjusted and rounded out. This is why there is interest in systematically measuring or evaluating the knowledge and skills around today, in order for them to still exist tomorrow. The many current contexts are to be considered systematically. It is a matter of familial, professional, institutional, economic, social, cultural, national and international contexts. These different contexts are made to serve entrepreneurial activity. In this regard, the precision provided by one of the entrepreneurs surveyed is significant: "*my future aspirations and challenges, which have to attract financing, are the Internet and international development*".

However, this requires a higher level of study and many skills: "*nowadays, you have to be a charismatic leader, a whiz with new technologies, bold and multitalented in the face of these technologies and different activities*". It is a question of cognitive acquisition, albeit expressed differently: "*in our generation, a lot of people have multiple skills because we are well trained, we get used to working in teams, working on projects, very early on; that might sound pretentious, but it's true*". Such demands are expedited by the use of digital technologies. It is this that gives the words *research, product, innovate* and *start-up* a chi index higher than 70[3].

The cognitive resources allow the technology entrepreneur to best combine the difficulties of the socioeconomic context with those of his or her organization. Moreover, part of the vocabulary has a positive tone and relates to the passion of the entrepreneurs for their professions. The following verbatim accounts are significant within this context ($chi^2 = 89$): "*I really like to take care of the business and I really like working as head of the company, passion for new technologies in general and for digital technology in particular, and the desire to create something, to create value, and this passion and these desires are constantly there; they are deep within me and guide me every day*". In this way, the technology entrepreneur is prepared by the fundamental factors of self-evaluation and the factors related to emotional and cognitive resources found in Table 3.1.

3 See footnote 2.

Dimensions of entrepreneurial identity	Verbatim perceptions of technology entrepreneurs in decreasing order of importance
Need to be: the entrepreneur's motivations and the key resources of the business	– New technologies and staff skills ($chi^2 = 89$) – Confidence in one's abilities or skills in digital technologies and strategic management ($chi^2 = 89$) – Passion, self-fulfillment, quality of life ($chi^2 = 86$) – Responsiveness to globalization ($chi^2 = 85$) – Independence (khi2 = 81), autonomy ($chi^2 = 79$), power ($chi^2 = 69$) – Financial resources and security ($chi^2 = 69$) – Balance between sport, culture and work ($chi^2 = 69$) – Adaptability/flexibility ($chi^2 = 69$) – Recruitment of skills given fears of missing one's targets and limited experiences ($chi^2 = 55$)
Need to know: knowledge, networks for support and sharing	– Openness to technological progress, innovation, creativity, and the international arena ($chi^2 = 89$) – Openness to new practices and cultures ($chi^2 = 87$) – Team spirit ($chi^2 = 87$) – Family support ($chi^2 = 87$) – Level of training ($chi^2 = 85$) – Search for financial and moral support ($chi^2 = 68$)
Need to be able: main skills	– Strategy ($chi^2 = 92$) – Challenge of internet, growth/international openness ($chi^2 = 89$) – Teamwork ($chi^2 = 89$) – Fear of failure or of not meeting expectations ($chi^2 = 85$) – Continuous responsiveness to creating loyalty among customers and their development ($chi^2 = 85$) – Performance obligations ($chi^2 = 84$) – Designing new projects ($chi^2 = 81$) – Delegation ($chi^2 = 70$) – Telecommuting and technological innovations ($chi^2 = 69$) – Experiences of spotters, advisers and associates ($chi^2 = 69$) – Establishing priorities ($chi^2 = 68$) – Numerous responsibilities ($chi^2 = 45$) – Stress and uncertainty ($chi^2 = 35$) – Long working hours ($chi^2 = 30$)

Table 3.1. *The most significant major occurrences in terms of* chi^2

The interpretative analysis of the content of this table is based on the three components of identity mentioned initially.

3.2.2.1. *Entrepreneurial motivations consistent with technological skills*

A number of motives feed the desire to launch, purchase or take over a business. There are two kinds of motivations identified by the entrepreneurs: the enthusiastic use of new technologies and the search for independence. "*My main motivation is working enthusiastically, with new technologies, with people that I've chosen on topics that I've chosen*". A greater number of motivational features were mentioned, including fulfillment of a dream, the search for a better quality of life, independence, power and autonomy. The start-up entrepreneur is also driven by the aim to prove their skills by using new technologies. They are the key resources of their organization.

3.2.2.2. *Knowledge underpinned by openness to creative values and technical innovation*

Within the framework of this study, the start-up founders have placed emphasis on certain values:

– openness to innovation, creativity, the international arena, change, competition;

– the search for new knowledge, and new cultures and practices.

These same values spark in the founders the desire to change the world, to be useful to society, and to be free and independent. One respondent explained that: "*I want to align my values with the prevailing values in my job as head of the business, and digital technologies are an enormous help to me in doing this*".

Moreover, it seems that innovation is perceived as a necessity in competitive fields that are often globalized and affected by technological changes. Control of these changes is facilitated by the level of study, type of training and being able to push oneself to the limit which is a real source of happiness and pleasure. The pleasure in working towards progress and the happiness of a job well done continually returns and lies in reconciling work and family, and personal and professional values.

3.2.2.3. *Need to be able to focus on technological and strategic skills*

The need to be able relies on technological skills and the ability to work in a team, but a more in-depth analysis of the body of information allows us to state that "strategic" skills are taken seriously. These are related to the operational dimensions of "delegating and sharing tasks", which were constantly mentioned during the three hours of the focus group. On the other hand, we can see that it is also a matter of projective skills that focus on digital technologies, creativity, academic training, the knowledge that one is surrounded by good people and well advised or supported, flexibility, and the search for novelty and for moral and financial family support.

The technology creator has followed the path to entrepreneurship by means of confidence in their abilities, moral and financial support from those close to them and bankers, as well as advice from spotters and associates. The support from one's immediate circle, expert advisers and the financial and technological communities appears to be indisputable. Such support seems to consolidate the components of the technology creator's entrepreneurial identity.

3.2.3. *Three main identity components to be considered when supporting the technology entrepreneur*

3.2.3.1. *Key characteristics: motivations, knowledge and experiences*

The results of our study show that the technology creator has a strong entrepreneurial attitude. This translates into a propensity to act independently, a will to innovate and take risks, and a tendency to be aggressive in terms of market opportunities [LUM 96]. Even while innovation remains the basis of all business creation, a start-up is based on knowledge used requiring a high level of qualifications and particular technical skills [NER 98, SEE 97].

At the level of our studies, these variables are the product of their life stories, university education and access to information made easier by openness to digital technologies [DAR 10]. In this area, they have all shown a real desire to make use of all of the techniques they can [SHA 04]. They also have the will to prove that a technology concept is worth being exploited [GAS 02]. Their motivations and ambitions are underpinned by steady belief in their project. Effectively, such beliefs are replaced by numerous imperatives. They are interlinked with training, creation, developing a business plan with a technical and financial feasibility study of the project, identifying relevant contacts, putting together an initial list of potential clients, etc. [SCH 00, DAH 00].

It goes without saying that the creation of a business founded on innovation and knowledge requires preparatory work that is more extensive and of a higher quality than equivalent work for non-innovative businesses. Moreover, the creation process is longer for technology businesses than for traditional businesses [ALL 03]. The extent of the level of knowledge can favorably impact this process.

One of the main differences between technology entrepreneurs and traditional entrepreneurs concerns the level of education. The creators of technology businesses are in fact more educated than the founders of traditional businesses [BOR 03]. With an average age of 36, our eight technology entrepreneurs are highly educated (5 years of higher education), but have very little professional experience [CHR 01, COL 01]. Significant professional experience contributes to the success of the business being launched [WIK 01]. This is especially true when there is a similarity between the new business and the organization for which the entrepreneur worked previously [BRÜ 96, COO 94]. In all cases, the majority of technology entrepreneurs have acquired or developed their technical and managerial knowledge by having previously worked in organizations in the same lines of business in which they go on to create their business [BER 00b, KOU 02]. Only two of the eight fall into this framework. Two others come from the spinning-off of family-run SMBs in the same line of business [LAS 05]. Technological spinning-off is a process through which a new business is created on the basis of one or several elements transferred from a parent or incubator organization. However, everything relies on an entrepreneurial team of two to five members [KAD 02].

The personality of the entrepreneur also plays an important role, as it has a significant influence on their organization. Within the entrepreneurial team, the entrepreneur puts themselves in a position of continuous learning; this encourages the implementation of another future structuring logic. Their perceptions change more quickly. They remain open to all innovations in order to avoid harming their business and frustrating their team. They end up creating a management system that makes the maximum possible number of people aware of their responsibilities and can involve remodeling the organization of the business, so that it can be as competitive as possible. There are collective and evolving skills and resources that are considered to be one of the main factors of their success [ASP 05, COL 05, FIL 03].

3.2.3.2. *The nature of resources specific to technology businesses*

The eight entrepreneurs surveyed stated the importance of a number of accessible resources on their paths. The importance of accessible resources for a technology business is decisive for its development and performance. The facts show that having an insufficient amount of suitable resources is one of the reasons that young businesses fail [ZHA 95]. One of the main objectives of a young business is to transform from an organization with few resources to an organization with a wealth of resources. The means that they lack are varied and related, *inter alia*, to capital, technology, and managerial and people skills.

Technological resources are the critical ingredient for the development of new products and services [SHA 01a, SHA 01b]. The degree of innovation of the technology used is one of the factors behind the success of new technology businesses [ASP 05]. However, technology resources develop with a greater degree of complexity than in traditional businesses. These developments require additional R&D efforts and unquestionable technical skills. Technology entrepreneurs, due to their training and experiences, are prepared to introduce and coordinate the technological resources that they need. Nevertheless, a number of works have underlined that these entrepreneurs lack other very important skills to ensure the success of the new business: skills in sales and marketing, management, finance and accounting [ALB 00, GAR 97]. This weakness was taken into consideration by our eight interviewees. They were aware of all the operational aspects of their business: the strategic vision, the mode of governance or management, the level of performance, the working methods, the marketing, the management of human resources, the logistics, etc.

They are also helped by an associate or a founding entrepreneurial team [RUE 03, SCH 00, TEA 03]. Thanks to their expert advice, they take other needs into account. It is a matter of qualified staff, of teams and premises adapted to the time required, to developing a new technology and to financing additional technologies. Indeed, the requirements of qualified staff drive them to adopt special recruitment and loyalty creation strategies [FOO 02]. Some entrepreneurs are reluctant to offer employees associate status. In this way, the entrepreneurial team takes the form of a true multidisciplinary group in terms of experience and knowledge [ASP 05, LAS 05].

It seems that the high cost of the basic technology used and of skilled human resources, the complexity of the market explored and the length of the development process for new technologies make the financial resource needs of technology entrepreneurs higher than those of traditional entrepreneurs [ALB 00, GAS 00]. Unfortunately, few entrepreneurs have access to venture capital or other sources of substantial funding [HEI 04, GOR 05, SHE 01]. It goes without saying that access to financial resources is one of the main difficulties encountered by entrepreneurs when creating a technology business [AUK 01, KOU 02, SHE 01].

In all of these cases, the most common method of management is reliant on logic based on proximity (spatial, hierarchical, operational, temporal, even familial) with embedded entities. The balance between private life and personal life, and the advice of support professionals (in terms of awakening and building the most suitable entrepreneurial skills) reinforces these proximity logics. It is this that Obrecht [OBR 16] translates in terms of proximities as the teaching of support has an impact on the creator's entrepreneurial identity through their routes, personality traits and personal abilities. The entrepreneurial spirit which is clear in strong convictions and in symbolic and irrational emotion is naturally affected. Self-knowledge, motivations and entrepreneurial skills develop under the influence of support with mutual trust.

3.2.3.3. *An environment based on mutual trust in operational and strategic networks*

Most often, technology entrepreneurs emphasize particular prominent features: a strong culture of new technologies, access to higher education, the search for moral support and financing, and the needs for advice and support. The search for true harmony between the entrepreneur's identity (in tune with new technologies) and the identity of the business created is equally normal [GAS 14]. They try to maintain such harmony by means of openness to new knowledge and practices [TRO 10]. They are aware of being unable to escape the positive or negative influence of the communities on which they are dependent [STA 15]. In fact, "*launching a business or an innovation is a collective process that begins in the business's environment and which is supported by mutual trust in immediate networks, stimulated by regional actors*" [JUL 03, p. 175]. They have specific needs that interact with the local economy in a particular way [SAP 03]. This is the case with research and development activities, and activities involving particular knowledge and expertise. The access and efficient use of such resources enables the very high level of intensity in competition in the global market to be managed most effectively [NOW 00].

This means that the technology entrepreneur needs institutional support at the regional and local level [FEL 12]. Institutional support can even encourage intersectoral groups at the regional or national level [DIA 11, PEZ 12, PEZ 13]. This is a way of expanding and consolidating collaborations that develop with a multitude of actors: financial, technological and commercial partners. Going forward, we know that establishing such partnerships determines the emergence, survival and performance of new technology organizations [NIC 03, SHA 02]. It is henceforth all part of an ecosystem that revolves around the technology entrepreneur to support them. The relationship between the entrepreneur's personal network, on the one hand, and the launch and growth of the business created, on the other, has clearly been demonstrated, as the connection between various networks is the main distinguishing criterion differentiating the innovation projects that succeed from those that fail.

Granovetter's analysis [GRA 83, GRA 85] on networks with weak signals and those with strong signals makes perfect sense. The former, little known and poorly understood by the entrepreneur, gives rise to original information through which later innovations are created. Those networks allow the possibilities of business to be uncovered. The latter, with which communication is easy and marked by shared trust between the components, plays a further role in validating the information received from elsewhere, and in providing support during the innovation process. The role of the trusted partnership was also highlighted by Fourcade [FOU 94] who pointed out that the network is essential to a small business that aims to acquire cutting edge technologies and benefit from a size effect. This is the case in social circles, sports clubs, or even family ties, which can play a role in creating an entrepreneur's reputation [BER 00a].

Networks facilitate relationships for trading resources and skills, open up new market opportunities and increase actions to remodel the environment for the good of the business [ZIM 02]. These trade relationships are themselves facilitated by technology incubators [DAM 08, DUB 14, MAT 14, MEN 11]. Incidentally, many technology incubators work with activities on innovations for social change [BOU 15, LÉV 14]. In so doing, they seek to truly accelerate the progress of young businesses towards innovative concepts of all kinds. The accelerator support is found in more accessible sources of funding, the logistics of accommodation or even *coworking* spaces, support for internationalization, etc.

It is within this system of partner-based management that we find the "club" of supported entrepreneurs. Post-creation support is therefore meant to be both individual and collective. The choice of partnership approach comes from the will of the entrepreneur who is constantly looking to change their organization. Alongside environmental pressure, the intentional activity of the manager is one of the main engines of organizational change. Cooperation thus becomes a tool for managing the causes of organizational inflexibility and inertia (unchanging cultural obstructions). Progressively, it leads the business away from being a passive or simply reactive entity in the face of environmental stimuli.

3.3. Conclusion

The first section of our chapter underlines the importance of the entrepreneurial identity of technology creators who are often immersed in a culture of flexibility, openness of spirit, innovation and responsiveness. The consideration of this need to have an identity is the basis of the scientific relevance of the object of this contribution: to improve the support for the process of creating technology businesses. The identity components reproduced (in a very synoptic manner) by our study are significant in this regard. These identity components are at the root of the entrepreneurial rationale applied by technology creators. These same identity components take the form of needs to be, to know and to be able, and show differences in meaning, the appropriation of which leads to approaches that gather resources of all kinds in aid of developing technological entrepreneurship [WOR 13]: market accessibility, human capital and workforce, investment and finance, a support system of advice or mentoring, start-up aid, education and training, and moral and cultural support.

From an individual standpoint, we have more or less determined the key success factors for a technology entrepreneur. We have seen, for example, that the entrepreneur's experience can have a significant impact on the success of their project. However, an entrepreneur must rely not only on their experience, but also particularly on their technology resources. Indeed, a technology entrepreneur is someone capable of accessing and coordinating technological, human and financial resources with a view to maximizing profitability [STA 14].

However, it is necessary to go into these empirical results in greater depth, as well as the proposed demonstration using epistemological, theoretical and practical arguments. Being able to understand the support of technology entrepreneurs and developers falls within this approach. This in-depth analysis must emphasize other identity components able to underpin the teaching of entrepreneurial support in response to similarities, differences and interconnections between generations of creators/buyers of technology businesses. The three areas of the interactive construction of entrepreneurial identity will be clarified in a functional manner: the predisposition area for recruiting knowledge and skills (individual and need to be), the appropriation area for acquiring and building knowledge and skills (need to know) and the fulfillment zone for the active and/or proactive implementation of skills in a technology project completed by the individual (need to be able to do).

3.4. Bibliography

[ALB 00] ALBERT P., "La création d'entreprises de haute technologie", in ALLEN K., STEARNS T.N. (ed.), *Histoire d'entreprendre – Les réalités de l'entrepreneuriat*, Éditions EMS, Paris, 2000.

[ALL 03] ALLEN K., STEARNS T., "Nascent high tech entrepreneurs: the who, where, when and why", in LIBECAP G.D. (ed.), *Advances in the Study of Entrepreneurship, Innovation and Economic Growth*, vol. 14, London, 2003.

[ALL 04] ALLEN K., STEARNS T., "Technology entrepreneurs", in GARTNER W.B., SHAVER K.G., REYNOLDS P.D. (eds), *Handbook of Entrepreneurial Dynamics – The Process of Business Creation*, Sage, London, 2004.

[ALV 01] ALVAREZ S.A., BUSENITZ L.W., "The entrepreneurship of resource-based theory", *Journal of Management*, vol. 27, pp. 755–775, 2001.

[ALV 02] ALVAREZ S.A., BARNEY J.B., "Resource-based theory and the entrepreneurial firm", in HITT M.A., IRELAND R.D., CAMP S.M. *et al.* (eds), *Strategic Entrepreneurship: Creating a New Mindset*, Blackwell Publishers, Oxford, 2002.

[ART 06] ARTHURS J.D., BUSENITZ L.W., "Dynamic capabilities and venture performance: the effects of venture capitalists", *Journal of Business Venturing*, vol. 21, no. 2, pp. 19–215, 2006.

[ASP 05] ASPELUND A., BERG-UTBY T., SKJEVDAL R., "Initial resources, influences on new venture survival: a longitudinal study of new technology-based firms", *Technovation*, vol. 25, pp. 133–1347, 2005.

[AUD 07] AUDRETSCH D.B., BÖNTE W., TAMVADA J.P., "Religion and entrepreneurship", *Social Science Research Network*, vol. 28, no. 6, pp. 774–789, 2007.

[AUK 01] AUKEN H., "Financing small technology-based companies: the relationship between familiarity with capital and ability to price and negotiate investment", *Journal of Small Business Management*, vol. 39, pp. 240–258, 2001.

[BEN 82] BENZECRI J.P., *Histoire et préhistoire de l'analyse des données*, Dunod, Paris, 1982.

[BER 00a] BERNASCONI M., MONSTED M., *Les Start-up High Tech*, Dunod, Paris, 2000.

[BER 00b] BERNASCONI M., "Les modèles de développement des entreprises high tech", in BERNASCONI M., MONSTED M. (eds), *Les start-up high tech: Création et développement des entreprises technologiques*, Dunod, Paris, 2000.

[BIR 02] BIRCH D., "Small business: slump? What slump?", interview in *Fortune*, Working papers of the Economic Studies and National Accounts Directorate (DESE), New York, pp. 3–41, December, 2002.

[BLA 13] BLANK S, DORL B., *Le Manuel du créateur de start-up: Etape par étape, bâtissez une entreprise formidable*, Les Editions Diateino, Paris, 2013.

[BOR 03] BORGES C., BERNASCONI M., FILION L.J., "La création des entreprises de hautes technologies – examen de la documentation", *Cahier de recherche* 2003-11, HEC Montreal, 2003.

[BOU 15] BOUCHARD M., TRUDELLEC J., BRIAND L. *et al.*, "A relational database to understand social innovation and its impact on social transformation", in NICHOLLS A., SIMON J., GABRIEL M. (eds), *New Frontiers in Social Innovation Research*, Palgrave Macmillan, London, 2015.

[BOU 94] BOUCHIKHI H., KIMBERLEY J., *Entrepreneurs et Gestionnaires*, Editions d'Organisation, Paris, 1994.

[BRO 94] BROWN D.D., "Discursive moments of identification", *Current Perspectives in Social Theory*, vol. 14, pp. 269–292, 1994.

[CAR 09] CARDON M.S., WINCENT J. *et al.*, "The nature and experience of entrepreneurial passion", *Academy of Management Review*, vol. 34, no. 3, pp. 511–532, 2009.

[CAS 96] CASTROGIOVANNI G., "Pre-start-up planning and the survival of new small businesses: theoretical linkages", *Journal of Management*, vol. 22, no. 6, pp. 801–822, 1996.

[CHR 01] CHRISTENSEN B., CHRISTIANSEN T., GOMBERT A.K. *et al.*, "Simple and robust method for estimation of the split between the oxidative pentose phosphate pathway and the Embden-Meyerhof-Parnas pathway in microorganisms", *Biotechnol. Bioeng.*, vol. 74, no. 6, pp. 517–523 PMID:1149421, 2001.

[COL 01] COLOMBO M., DELMASTRO M., "Technology-based entrepreneurs: does internet make a difference?", *Small Business Economics*, vol. 16, pp. 177–190, 2001.

[COL 05] COLOMBO M., GRILLI L., "Founders' human capital and the growth of new technology-based firms: a competence-based view", *Research Policy*, vol. 34, pp. 795–816, 2005.

[COO 94] COOPER A.C., GIMENO-GASCON F.J., WOO C.Y., "Initial human and financial capital as predictors of new venture performance", *Journal of Business Venturing*, vol. 9, no. 5, pp. 371–395, 1994.

[DAH 00] DAHLQVIST J., DAVIDSSON P., WIKLUND J., "Initial conditions as predictors of new venture performance: a replication and extension of the cooper *et al.* study", *Enterprise & Innovation Management Studies*, vol. 1, no. 1, pp. 1–17, 2000.

[DAM 08] DAMBRON P., *Les clusters en France, pourquoi les pôles de compétitivité?*, L'Harmattan, Paris, 2008.

[DAN 10] DANA L.P., *Entrepreneurship and Religion*, Edward Elgar, Cheltenham, 2010.

[DAR 10] DARMON M., *La socialisation: Domaines et approches*, Armand Colin, Paris, 2010.

[DEC 99] DECOURT P., L'identité et la perte, *Revue française de psychanalyse*, vol. 63, no. 4, pp. 1153–1164, 1999.

[DIA 11] DIAMANTOPOULOS M., "Cooperative development gap in Québec and Saskatchewan 1980 to 2010: a tale of two movements", *Canadian Journal of Nonprofit and Social Economy Research*, vol. 2, no. 2, pp. 6–24, 2011.

[DOK 14] DOKOU KOKOU G.A., "Identité territoriale, proximités locales et capacités entrepreneuriales", in KOKOU DOKOU G.A. (ed.), *Acteurs territoriaux, identités et capacités entrepreneuriales*, L'Harmattan, Paris, 2014.

[DUB 14] DUBEUX A.M., "Education Populaire et Economie Solidaire : la logique de FASAGES", *RIUESS 2014 – XIVe Rencontres du Réseau Inter-Universitaire de l'Economie Sociale et Solidaire*, p. 19, 21–23 May, Lille, 2014.

[EAS 14] EASTMAN J., IYER R., LIAO-TROTH S. *et al.*, "The role of involvement on millennials' mobile technology behaviors: the moderating impact of status consumption, innovation, and opinion leadership", *Journal of Marketing Theory and Practice*, vol. 22, no. 4, pp. 455–470, 2014.

[ERI 68] ERIKSON E.H., *Identity Youth and Crisis*, W.W. Norton, New York, 1968.

[FEL 12] FELD B., *Start-Up Communities: Building an Entrepreneurial Ecosystem in your City*, John Wiley & Sons Inc., New York, 2012.

[FIL 03] FILION L.J., BORGES C.V., BERNASCONI M., "La création des entreprises de hautes technologies: examen de la documentation", *Cahier de recherche 2003–11*, p. 39, July 2003.

[FOO 02] FOO M.D., TAN H.H., "Critical incidents in high-tech start-ups: a Singapore study", in PHAN P. (ed.), *Technological Entrepreneurship*, 2002.

[FOU 94] FOURCADE C., "Localisation versus globalisation: les stratégies résiliaires des petites enterprises", *ICSB World Conférence*, Strasbourg, 6–8 October, 1994.

[GAR 97] GARTNER W., STARR J., BHAT S., "Predicting new venture survival: an analysis of anatomy of a start-up", *Journal of Business Venturing*, vol. 14, no. 2, pp. 215–232, 1997.

[GAS 00] GASIGLIA C., GUEYE S., PISTRE N., "La valorisation, enjeu central de la logique financière des start-up", in BERNASCONI M., MONSTED M. (eds), *Les start-up high tech: Création et développement des entreprises technologiques*, Dunod, Paris, 2000.

[GAS 02] GASSE Y., *Les entrepreneurs technologiques: le profil des chercheurs-entrepreneurs*, Miméo, 2002.

[GAS 14] GASSE Y., TREMBLAY M., Etude sur les entrepreneurs et les repreneurs québécois de générations X et Y, FSA, Université Laval, Quebec, 2014.

[GHI 78] GHIGLIONE R., MATALON B., *Les enquêtes sociologiques, théories et pratiques*, Armand Colin, Paris, 1978.

[GHI 85] GHIGLIONE R., MATALON B., BACRI N., *Les dires analysés, l'analyse propositionnelle du discours*, PUF, Paris, 1985.

[GOR 05] GORMAN G., ROSA P.J., FASERUCK A., "Institutional lending to knowledge-based business", *Journal of Business Venturing*, vol. 20, pp. 793–819, 2005.

[GRA 83] GRANOVETTER M., "The strength of weak ties: network theory revisited", *Sociological Theory*, vol. 1, pp. 201–233, 1983.

[GRA 85] GRANOVETTER M., "Economic action and social structure: the problem of embeddedness", *American Journal of Sociology*, vol. 91, no. 3, pp. 481–510, 1985.

[HEI 04] HEIRMAN A., CLARYSSE B., "How and why do research-based start-ups differ at founding? A resource-based configurational perspective", *Journal of Technology Transfer*, vol. 29, pp. 247–268, 2004.

[HIT 03] HITLIN S., "Values at the core of personal identity: drawing links between two theories of self", *Social Psychology Quarterly*, vol. 66, no. 2, pp. 118–137, 2003.

[HOA 10] HOANG H., GIMENO J., "Becoming a founder: how founder role identity affects entrepreneurial transitions and persistence in founding", *Journal of Business Venturing*, vol. 25, no. 1, pp. 41–53, 2010.

[HOG 00] HOGG M.A., TERRY D.J., "Social identity and self-categorization processes in organizational contexts", *Academy of Management Review*, vol. 25, pp. 121–140, 2000.

[JUL 03] JULIEN P.A., "Innovation et entrepreneuriat", in MUSTAR P., PENAN H. (eds), *Encyclopédie de l'innovation*, Economica, Paris, 2003.

[KAD 02] KADJI-YOUALEU C., FILION L.J., "Essaimage technologique: Examen de la documentation", *Cahier de recherche 2002-14*, HEC Montreal, 2002.

[KIM 79] KIMBERLY J., "Issues in the creation of organizations: initiation, innovation and institutionalization", *Academy of Management Journal*, vol. 22, pp. 437–457, 1979.

[KOU 02] KOURILSKY M.L., WALSTAD W.B., "The early environment and schooling experiences of high-technology entrepreneurs: insights for entrepreneurship education", *International Journal of Entrepreneurship Education*, vol. 1, pp. 87–106, 2002.

[LAS 05] LASCH F., LE ROY F., YAMI S., "Les déterminants de la survie, de la croissance des start-up TIC", *Revue française de gestion*, vol. 31, no. 155, pp. 37–56, 2005.

[LEB 00] LEBART L., MORINEAU A., PIRON M., *Statistique Exploratoire Multidimensionnelle*, Dunod, Paris, 2000.

[LEV 14] LEVEQUE M., "Psychosurgery: New Techniques for Brain Disorders", Springer, 2014.

[LIT 98] LITTUNEN H., STORHAMMAR E., NENONEN T., "The survival of firms over the critical first 3 years and the local environment", *Entrepreneurship and Regional Development*, vol 10, pp. 189–202, 1998.

[LUM 96] LUMPKIN G.T., DESS G.G., "Clarifying the entrepreneurial orientation construct and linking it to performance", *Academy of Management Review*, vol. 21, no. 1, pp. 135–172, 1996.

[MAR 97] MARC E., "L'identité personnelle", *Sciences Humaines, Hors-Série: Identité, Identités*, vol. 15, pp. 6–8, 1997.

[MAT 14] MATRAY M., POISAT J., "Les PTCE, une dynamique citoyenne et institutionnelle d'innovations socials", *Cahiers du CIRTES* Hors-Série 4, Presses Universitaires Louvain, Louvain, 2014.

[MEN 11] MENU S., "Les Pôles de compétitivité, un nouveau pilotage de la politique industrielle?" *Politiques et Management Public*, Bilan en Ile-de-France, vol. 28, no. 1, 2011.

[MUC 74] MUCCHIELLI R., *L'analyse de contenu*, E.M.E., Paris, 1974.

[MUR 07] MURNIEKS C.Y., MOSAKOWSKI E.M., "Who am I? Looking inside the entrepreneurial identity", *Frontiers of Entrepreneurship Research Proceedings*, Babson College Entrepreneurship Research Conference (BCERC), Madrid, Spain, pp. 7–9, 2007.

[NER 98] NERLINGER E., "Standorte und Entwicklung junger innovativer Unternehmen: Empirische Ergebnisse für West-Deutschland, Baden-Baden", *Schriftenreihe des Zentrum für Europäische Wirtschaftsforschung* (ZEW), vol. 27, no. 3, pp. 111–132, 1998.

[NIC 03] NICOLAOU N., BIRLEY S., "Academic networks in a trichotomous categorisation of university spinouts", *Journal of Business Venturing*, vol. 18, no. 3, pp. 333–359, 2003.

[NOW 00] NOWAK M.J., GRANTHAM C.E., "The virtual incubator: managing human capital in the software industry", *Research Policy*, vol. 29, pp. 125–134, 2000.

[OBR 16] OBRECHT J.J., "Sustainable entrepreneurship education: a new field for research in step with the 'effectual entrepreneur'", *International Journal of Entrepreneurship & Small Business*, vol. 29, no. 1, pp. 83–102, 2016.

[PEZ 12] PEZZINI E., "La capacité d'innovation et d'exploration des nouveaux secteurs coopératifs. Le cas italien", *International Summit of Cooperatives*, Quebec, 2012.

[PEZ 13] PEZZINI E., "Les associations des coopératives. Potentialités, défis, modèles d'organisations", *Conférence internationale sur la recherche en économie sociale du CIRIEC International*, Anvers, 24–26 October 2013.

[REI 98] REINERT M., "Processus catégorique et co-construction des sujets dans différents récits", *Linguistique et Psychanalyse, Colloque de Cerisy-La-Salle*, September, 1998.

[REY 02] REYNOLDS P., BYGRAVE W., AUTIO E. *et al.*, Global Entrepreneuship Monitor, 2002 Executive Report (GEM 2002 Summary Report, Global Entrepreneuship Monitor), Washington, pp. 10–39, 2002.

[RIC 90] RICOEUR P., *Soi-même comme un autre*, Le Seuil, Paris, 1990.

[RIE 11] RIES E., *The Lean Start-Up: How Today's Entrepreneurs Use Continuous Innovation to Create Radically Successful Businesses*, Crown Publishing Group, New York, 2011.

[RUA 04] RUANO-BORBALAN J.C., "L'identité, une ressource pour l'action", in HALPERN C., RUANO-BORBALAN J.C. (eds), *Identité(s). L'individu. Le groupe. La société*, Editions Sciences Humaines, Paris, 2004.

[RUE 03] RUEF M., ALDRICH H.E., CARTER N.M., "Homophily, strong ties, and isolation among U.S. entrepreneurs", *American Sociological Review*, vol. 68, pp. 195–222, 2003.

[SAP 03] SAPIENZA H.J., AUTIO E., ZAHRA S., "Effects of internationalization on young firms' prospects for survival and growth", *Academy of Management Best Conference Paper,* 63, ENT G1–7, 2003.

[SAR 08] SARASVATHY S.D., *Effectuation: Elements of Entrepreneurial Expertise*, Edward Elgar, Cheltenham, 2008.

[SCH 00] SCHUTJENS V., WEVER E., "Determinants of new firm success", *Papers of Regional Science*, vol. 79, pp. 135–159, 2000.

[SEE 97] SEEGER H., "Ex-post bewertung der technologie und gründerzentren durch die erfolgreich ausgezogenen unternehmen und analyse der einzel undregionalwirtschaftlichen effekte", Münster, *Hannoversche Geographische Arbeiten*, vol. 53, no. 4, pp. 1–15, 1997.

[SHA 01a] SHANE S., "Technology regimes and new firm formation", *Management Science*, vol. 47, no. 9, pp. 1173–1190, 2001.

[SHA 01b] SHANE S., "Technological opportunities and new firm creation", *Management Science*, vol. 47, no. 2, pp. 205–220, 2001.

[SHA 02] SHANE S., STUART T., "Organizational endowments and the performance of university start-ups", *Management Science*, vol. 48, no. 1, pp. 154–170, 2002.

[SHA 04] SHANE S., "Encouraging university entrepreneurship? The effect of the Baby-Dole act on university patenting in the United States", *Journal of Business Venturing*, vol. 19, pp. 127–151, 2004.

[SHE 01] SHEPHERD D.A., ZACHARAKIS A., "The venture capitalist-entrepreneur relationship: control, trust and confidence in cooperative behaviour", *Venture Capital: An International Journal of Entrepreneurial Finance*, vol. 3, pp. 129–149, 2001.

[SOL 00] SOLMOSSY E., "Entrepreneurial dimensions: the relationship of individual, venture and environmental factors to success", *Entrepreneurship, Theory & Practice*, pp. 79–80, 2000.

[STA 14] STAM E., *Entrepreneurial Ecosystems and Regional Policy: A Sympathetic Critique*, Ultrecht School of Economics, Koopmans Research Institute, pp. 15–37, 2014.

[STA 15] STAM E. "Entrepreneurial Ecosystems and Regional Policy: A Sympathetic Critique", *European Planning Studies*, vol. 23, no. 9, pp. 1759–1769, 2015.

[STR 92] STRAUSS A., *Miroirs et masques*, Métailié, Paris, 1992.

[TAP 97] TAP P., "Marquer sa différence", *Sciences Humaines, Hors-Série: Identité, Identités*, vol. 15, pp. 9–10, 1997.

[TEA 03] TEAL E.J., HOFER C.W., "New venture success: strategy, industry structure, and the founding entrepreneurial team", *Journal of Private Equity*, vol. Fall, pp. 38–51, 2003.

[TEE 07] TEECE D.J., "Explicating dynamic capabilities: the nature and microfoundations of (sustainable) enterprise performance", *Strategic Management Journal*, vol. 28, pp. 1319–1350, 2007.

[TOD 03] TODTLING F., WANZENBOCK H., "Regional difference in structural caracteristics of start-up", *Entrepreneurship and Regional Development*, vol. 15, pp. 351–370, 2003.

[TRO 10] TROPE Y., LIBERMAN N., "Construal-level theory of psychological distance", *Psychological Review*, vol. 117, no. 2, 2010.

[VIB 11] VIBERT P., "L'identité: Un nouveau chantier théorique?", in LAZZAROTTI O., OLAGNIER P.-J. (eds) *L'identité. Entre Ineffable et effroyable*, Armand Colin, Paris, 2011.

[WIK 01] WIKLUND J., SHEPHERD C., "The sustainability of the entrepreneurial orientation-performance relationship", *Entrepreneurship Theory & Practice*, vol. Fall, pp. 37–48, 2001.

[WOR 13] WORLD ECONOMIC FORUM, "Entrepreneurial ecosystems around the globe and company growth dynamics", *Report Summary for the Annual Meeting of the New Champions 2013*, Davos, 2013.

[ZAH 06] ZAHRA S.A., SAPIENZA H.J., DAVIDSSON P., "Entrepreneurship and Dynamic Capabilities: A Review, Model and Research Agenda", *Journal of Management Studies*, vol. 43, no. 4, pp. 917–955, 2006.

[ZHA 95] ZHAO L., ARAM J.D., "Networking and growth of young technology-intensive ventures in China", *Journal of Business Venturing*, vol. 10, no. 5, pp. 349–370, 1995.

[ZIM 02] ZIMMERMAN M.A., ZEITZ G.Z., "Beyond survival: achieving new venture growth by building legitimacy", *Academy of Management Review*, vol. 27, no. 3, pp. 414–431, 2002.

4

The Importance of Entrepreneurial Creativity

Creativity is a field that only began to be studied after innovation, and whose shapes and content are still indistinct. Creativity is at once both a process and an outcome of this process which allows it to produce ideas. When it does not concern products and services intended for customers, it is described as organizational creativity. Woodman, Sawyer and Griffin [WOO 93] considered organizational creativity to be a subsection of innovation. For Amabile [AMA 88], creativity is at the root of innovation by producing new and useful ideas, and can be considered its forerunner. Through his approach to leadership, Deschamps [DES 03] specifies that stimulating, guiding and supporting innovation requires a very particular emphasis to be placed on the development of current and future skills which are necessary to the given business. Among these skills are the creativity, know-how, values, state of mind and behaviors that create innovations. Innovation is often approached through its end result, while creativity often seems to represent the origin or cause. Incidentally, some creators do concentrate on the process to come, which can encourage innovation and an interest in their capacity to innovate [PAR 15].

Creativity is thus presented as a dynamic capability that stimulates innovative activity and leads the organization to adopt risky behaviors and changes in its practices. Despite this, it is the topic of innovation that has been the subject of a significant number of academic studies in recent decades. Moreover, the vast majority of research focused on technological

---

Chapter written by Marc JAILLOT.

innovation and, more recently, only on organization-driven innovation [DAM 12]. Crossman and Apaydin [CRO 10] showed that half of the research articles do not specify the type of innovation studied and, among those that do, only 3% are about organizational innovation. In another literature study based on seven journals and 342 articles published between 1992 and 2010, Keupp *et al.* [KEU 11] identified only 25 articles that included organizational innovation. Studies on innovation therefore appear to focus on technological innovation, and theories and practices are thus naturally also based on this approach [DAM 12] and provide even less for questions on the organizational aspect. Similarly, creativity and innovation are often addressed together to describe the product-led innovation that allows a business to keep a "star" product, but there are much fewer of these as it involves successful combinations of multitudes of ideas that achieve entrepreneurial success, and is therefore based on a diffuse kind of innovation rather than on the one best idea. Yet, this creative approach, whose object is not necessarily a technological innovation, can be difficult to imitate, and therefore constitute protection of a competitive advantage, even though industrial protection in the form of a patent has not been established, as is often the case in small businesses.

The aim of this chapter is to question the concept of entrepreneurial creativity by comparing it to the idea of innovation in order to determine its place and importance within small businesses. The first section will clarify the concept of entrepreneurial creativity and the chosen context. We will then indicate the elements that highlight the importance of creativity for entrepreneurs with small businesses by relying on practical experiences that will shed light on our intentions and pave the way for a discussion in the final section.

4.1. Creativity, innovation and entrepreneurial context

4.1.1. *Entrepreneurial creativity*

The creativity that we intend to study in this chapter is the creativity particular to entrepreneurship. Entrepreneurship is described as a process that leads to the creation of a new organization [GAR 95] and which is driven by one or several individuals who are working together on the project [VER 03]. These individuals often undertake such activity when they find a business opportunity. The entrepreneur therefore becomes an individual who senses an opportunity and creates an organization to pursue it [BYG 91]. Verstraete and Fayolle [VER 05] have proposed a depiction of entrepreneurship using four paradigms: the creation of an organization, the business opportunity, the innovation and the creation of value. The concept of entrepreneurship often refers to the creation of a business. However, the literature allows us to define it more broadly as the phenomenon in which an entrepreneur develops, whether they are creating a project, taking one over or developing a project unique to their organization. The intrapreneurial process is an integral part of entrepreneurship [CAR 97]. The author explains that entrepreneurs and intrapreneurs are distinguished by the fact that the intrapreneur activates the entrepreneurial process in an existing organization, such as by means of a new project, while the entrepreneur is involved in the launch or development stage of a newly created business.

The generic term "entrepreneur" encompasses two types. Their profiles are similar and it is primarily their operational context that differentiates them. The interest in engaging in the entrepreneurial aspect of creativity lies in the fact that this leads to harnessing the entire creative phenomenon by means of the entrepreneur who holds the keys to developing the business. Entrepreneurs, and particularly entrepreneurs with small- and medium-sized businesses (SMBs), seek creativity not only when they create a new product or process, but also, and perhaps especially, when they organize their value creation system. For example, Carrier [CAR 97] showed that an entrepreneur must often have evidence of creativity to find the capital required for their activity or to convince investors and clients of the benefit of their product or service. The creative skills of individuals are based on certain characteristics, such as risk-taking, attitudes to uncertainty or failure and independence, all of which are also attributes of entrepreneurs [BIS 11]. Creativity is even considered part of the entrepreneurial spirit [TSA 14] and plays an important role throughout the entrepreneurial process [NYS 93].

4.1.2. *The entrepreneurial context of small businesses*

The entrepreneurial context is important and being interested in a small business raises two questions: what is a small business and why would one be interested in this type of organization? Our decision to define and study small businesses is, in itself, of interest: by choosing the lens that creates the shape of the population studied, we predetermine the framework of the analysis and, therefore, the results. Most research gives the impression that the choice is not incumbent upon the researcher, but rather on the institutions which have determined the criteria for dividing the business population, primarily at the national level, and then the community and international levels. These criteria are reasonably quantitative as it is easier to set them up for data collection and analysis. While there is a strong tendency to homogenize criteria and establish the relevance of research, such as through the categorization proposed by the French Law on Modernization of the Economy (Art. 51 LME, 2008–1354), we can see that there are different visions around the world about what constitutes a small business. Incidentally, other types allow us to split up businesses differently [JUL 90]. Nevertheless, numerous authors have shown that size is certainly the factor that most impacts and limits the structure of an organization [MIN 82, DES 92], even if some have written that there are no "laws" on the link between size and organizational characteristics [HAL 97]. The division we have chosen is based on the criterion of size, but we still wish to remain aware of two risks of potential errors: wrongly accepting the assumption that small businesses have specific features, and mistakenly rejecting microbusinesses or middle-market companies on the basis that they are of a different size [TOR 97]. We also wish to guard against a second angle that consists of confusing small businesses that are truly autonomous under French law (decree 2008–1354) with those that are legally independent but owned by larger companies. This is the case with legal units, for instance, which are legally independent structures but can sometimes conceal less independent kinds of business under the control of large groups.

We will conclude our journey through the segment on targeted businesses with the reasons that we currently have for studying small businesses which generally fall within the social role given to businesses and entrepreneurship: the capacity to create jobs and added value. The development of entrepreneurship consequently appears in numerous institutional programs, often as a potential contributing element to development or economic recovery. The stated objectives are most often found at the level of employment and wealth redistribution, which is often justified by extensive research establishing the correlation between entrepreneurship and growth [HOW 98, AUD 01]. The World Bank, the OECD and the European Union, as well as non-governmental organizations and business networks, encourage the promotion of entrepreneurship as a fundamental instrument of growth, development, social inclusion and eliminating poverty. The European Commission, for instance, proposed the Entrepreneurship 2020 Action Plan which, as part of the strategy, is laying the groundwork for "growth and competitiveness that will be smart, sustainable and inclusive" (Commission to the European Parliament, 2013). Concrete measures are appearing in certain areas, including education policy, for instance. Work has shown that the desire to create a business can be influenced in advance during training periods. The creation in France of the "student entrepreneur" as a nationally recognized status and of a related diploma [CIR 13] or the use of business simulation games [TOU 10] are examples of the responses aiming to inspire the desire to create a business among young academics or students.

Being interested in small business involves studying 99.8% of French small- and medium-sized businesses (including microbusinesses), which represent 46% of the workforce, or 6 million full-time equivalent jobs [INS 14]. If we focus solely on SMBs without taking microbusinesses into account, it is 4% of businesses that employ almost 28% of workers, or 3.7 million people. The entrepreneurs behind microbusinesses and SMBs are therefore in control of a powerful tool in terms of jobs. As regards other categories, the 243 large businesses and 5,223 middle-market companies employ over half of employees, while representing only 0.15% of businesses of all sizes. The lens of the Pareto principle had thus far led us to optimize activities towards the few businesses with the most jobs. However, on the one hand, these large businesses often have the human and financial resources to make progress and innovate, and, on the other hand, it would be a shame to leave out businesses that employ over 6 million people. These are also the businesses that have the potential to become future greats, even if, regardless of the activity, the launch stage in the early years is often carried out as a reduced organization with fewer people. In fact, 95% of businesses created actually start without any employees, and those that have employees from the beginning have an average of 2.7 people [INS 14]. In spite of this, for almost 10 years, in both France and the European Union, SMBs have been behind the creation of over 80% of new jobs, half of which have been in businesses with fewer than 20 employees. SMBs (including microbusinesses) employ the majority of their workers in customer-facing services, as well as in specialized scientific and technical activities. Microbusinesses employ 70% of their paid staff in service industries. Overall, SMBs that are not microbusinesses have thirty employees on average. The majority (60%) have less than 20 employees and only 12% reach or exceed 50 employees. The activities occurring within SMBs are more varied than for other categories, but they are over-represented in construction and traditional industries. In the construction sector, which still includes around 10 large businesses, 71% of paid workers are employed in SMBs [INS 12].

4.1.3. *The reality of innovation in small businesses*

Through reading reports, we have discovered that some characteristics are specific to small businesses and confirm that they can be more than a particular field of observation, becoming an "*object of study in their own right, ontologically established*" [MAR 03]. First, the distribution of the creation of added value shows that small and very small businesses create less added value per employee and per year than middle-market companies and large businesses [INS 16]. Large businesses are the bigger suppliers of added value with 96.4 K€/employee. Middle-market companies are at 80.6 K€/employee, while other smaller businesses, SMBs and microbusinesses generate 74.4 K€/employee. The gap between SMBs and microbusinesses is also notable as the latter have a greater capability to create added value with 88.6 K€/employee compared to 65.0 K€/employee for SMBs, for a similar turnover per employee. SMBs therefore generate less added value per employee. These figures are the well-established averages, but such differences between SMBs and microbusinesses encourage us to carry out a differentiated analysis.

We have also been able to observe that small businesses are not exporters. Over 56% of added value is produced by around 5,200 businesses (large businesses and middle-market companies), which carry out 85% of exports. The most concentrated sector is the finance and insurance sector, in which around 30 large businesses employ 78% of workers. The French publication *Insee Première* [INS 14] highlighted the fact that it is exporting businesses that are more innovative and that exporting businesses are not SMBs: only 15.5% of exportation in France is carried out by small businesses (SMBs including microbusinesses).

The third point that characterizes the small business is the fact that innovation does not seem to be a central issue, in the sense of searching for something new by means of a developed process. Regardless of the level of research expenditure, the ability to innovate or the filing of a patent, the bottom line is that this is an area reserved for large businesses and middle-market companies. In 2014, 12,567 patent applications were filed with the National Intellectual Property Institute (INPI) by French businesses for national protection, and of those 12,567 patent applications, 6,715 (or 53.4%) were made by 20 businesses that are all large businesses [INP 15]. More broadly, research expenditure in France is highly concentrated in certain areas of research. In 2013, industrial areas received 79% of national expenditure on business enterprise research and development (BERD). Among these, the automotive industry, aircraft and spacecraft construction, as well as the pharmaceutical industry together receive 34% of all expenditure. The automotive industry, with 13% of the BERD total, is the top area of research and it is notable that industries represent only 8.10% of businesses (companies) and only 4.84% of businesses created [INS 14]. Beyond industry and construction, 8 in 10 businesses carry out service activities and, therefore, do not belong to the areas deeply involved in innovation. Until now, these service activities have been rarely studied, even though they are numerous and are major sources of value creation in our largely service-based economies [DES 12]. This result may, however, have to be nuanced depending on the chosen filter. Saying that small businesses are or are not innovative on the mere basis of their projects is a difficult task and it can allow significant differences to emerge depending on whether the business is taken as a legal entity or in the context of the 2008 French decree. BPI France, in its report on SMBs, recommended criteria related to its research and development projects, and the fact that it had issued at least one patent to the INPI in the previous two years [BPI 15]. Studies on innovative businesses remain complicated, being split initially between the subject approach, based on behavior and the business's innovation activities, and the object approach which is based on the number and features of the innovations produced [EUR 05]. Beyond large businesses and middle-market companies, the types of businesses that innovate the most are either SMBs with over 10 employees and more than 20 years of experience, or new young microbusinesses with less than five years of experience. In comparison to external sources of innovation, a substitute for internal development, Archibugi, Cesaratto and Sirilli [ARC 91] established a relationship between the size of the business and the importance granted to each type of source: the acquisition of patents and expertise plays a more important role in large businesses, whereas intermediate goods acquired as an element of innovation are the preferred medium of smaller businesses.

Lastly, an important point in terms of specific features is that innovation correlates with context. A more informed large business defines its area of originality in relation to the wider world and global markets, while an SMB is less able to completely understand newness and defines innovation in a relative sense in relation to a narrower but familiar context, a more localized market. Innovation can be a rearrangement of old ideas as long as the idea is seen as new by the people involved, as it may seem to others like an imitation of something that might already exist, and to take ideas, individuals, relationships and context into consideration [VAN 86]. Innovation resources can therefore easily be found at the investment level (in physical equipment, for example), which becomes innovation for the business in relation to its usual level of use (as well as its market).

We can thus raise the question of interest in making debates on innovation central as a means of competition, as a very low percentage of small businesses seem able to make a complete attempt and the sectors of activity involved in technological innovation only affect a tiny number of companies that generally have the resources and structures to guarantee the correct process. It is clear that they know how to communicate, explain, create and support businesses that have already mastered this process, and that they are increasingly interested in those who are able to carry out the process and rely on innovation support structures. However, what can be done as regards innovation for the 99% of remaining businesses?

4.2. What does the economic reality teach us?

The study of small businesses is interesting as it allows us to understand a method of development that does not necessarily involve either innovation or exports, but which shows a lesser capacity of value creation than businesses in other categories and whose sources of innovation vary depending on context. We suggest going into greater depth in our analysis of small businesses through the lens of creativity, which can explain part of the operating mode. The reality on the ground reveals aspects that cannot be found in the literature. Our research objective is to explain and understand how creativity can be an operating mode specific to small businesses. This involves identifying the kinds of creativity produced by the entrepreneur and collecting information on the creative approach.

Our knowledge acquisition occurs through an exploratory approach using semi-structured qualitative interviews facilitating an open discussion that does not limit the proposals received and allows a methodical opportunism [GIR 89]. The data obtained could be both quantitative and qualitative. The questioning was carried out in the entrepreneur's workplace in order to be able, as far as possible, to confirm the information extracted from the interview by closely observing the environment. David [DAV 04] actually shows the need to describe the context of a case study and specify the context to which it refers in order to make it easier to guarantee the validity and reliability of results. The choice is to keep observer participation to a minimum.

The study of creativity in smaller businesses nevertheless remains difficult in the sense that it often reduces us to studying the psychology of the entrepreneur, when we would rather be developing an explanation of the entrepreneurial situations encountered. Out approach therefore does not consist of focusing on the entrepreneurial actor and their traits, but on having a vision of relationships between individuals and their context as an entrepreneurial situation [SCH 09a]. We have addressed the problems of small businesses using the depictions of entrepreneurial actors within a research framework based on a real and tested approach [SCH 09b]. The main difficulties lie in reaching agreement on an observation based on an honest and sincere relationship, and in correctly describing that which our observations show us. Privacy can be an obstacle to accessing data, particularly on issues related to competitive advantages or accounting data. For that reason, we have adopted a progressive observation approach, as recommended by Lee [LEE 93]. A semi-structured interview guide has been developed in order to control the outline of the discussion. Being immersed in the activity while collecting data from the entrepreneur at the same time allowed us to effectively corroborate the conversation with the situation.

Three quite specific creativity situations have been identified and can constitute different types. The first type is based on a relatively traditional situation of economic arbitration using the production factors of work and capital. When these are combined with a context unique to each client or each product, the number of potential combinations very quickly becomes important. This mission to organize, dedicated to entrepreneurs with small businesses, can very quickly lead to the appearance of solutions that have never been used and are considered new by the business. The activity therefore moves away from being a repetitive and routine process, and requires knowledge and/or expertise on the part of the entrepreneur setting up an original productive organization. We therefore do not refer to a new process when it is a question of a range of possibilities leading to the conclusion of the mission. It is not possible to simply talk about innovation, but rather of the creative application of the elements necessary to carrying out an activity. The entrepreneur, constantly focusing on the overall management of their activity, concentrates at that moment in time on the production operation, like the head of a large entity. This can be called practice-led creativity, a form of organizational creativity that is not necessarily reliant on an intentional innovation process and which takes problem-solving methods into the field of theoretical application.

The second type is built on the desire of the entrepreneur to improve the production process. This improvement in production is targeted using traditional criteria for reducing costs and/or maintaining quality in accordance with specifications. It can also aim for improvement in order to deliver a higher value. The entrepreneur, fortuitously or unintentionally, recovers information during an exploratory stage and then develops the idea of using new knowledge within their business. They alter their manufacturing process, thereby creating something that is new to them, but which is without real innovation for the market. This appears very close to being an innovative product or process, without actually being protected by a patent. This solution is new for businesses, and is lacking real certainty as regards its reliability or durability. The creativity described here does not correspond to the traditional product innovation process: no research and development budget, no patent application. The entrepreneur generally only has the keys to this process as this type of creativity is particular to small business entrepreneurship, where the culture of secrecy only exists in a centralized way in operationalizing activity.

The third type of creativity can arise when there are urgent problems to be solved during the realization of an activity. The urgency is related to the fact that a complete or partial production chain is immobilized or blocked, leading to fixed costs for non-usage of the immobilized asset and employees, as well as for not providing value. The creativity therefore put in play is necessary as the problem has not been anticipated and therefore not included in a resolution procedure. This situation could be perceived in the same way as the situation described above, as a kind of problem solving. In fact, the creativity produced in this case is considered useful as the problem is resolved even though it had not been predicted. The type of urgency nonetheless leads the entrepreneur to make decisions in a very restricted context in terms of time and situation, and requires a powerful yet spontaneous creativity which is limited to solving the problem. The creativity can therefore take an unexpected form, in the Jugaad style, for instance, as the entrepreneur's sole concern at the time is to overcome the problem. It can even be useless or irrelevant in cases where formalization provides for a process, but it is one which has not been activated or respected. This creativity can therefore be declared potentially superfluous as it brings nothing to the activity. In fact, the problem should not have appeared or should have been addressed according to an established process, but, in light of the numerous potential combinations and the difficulty for an entrepreneur to understand all of the possible scenarios, it is likely that this type of situation will often occur. This will be particularly true in small businesses where the centralized form of management facilitates neither the distribution of information nor the thorough development of procedures.

The three aspects identified as part of the phenomenon of entrepreneurial creativity allow us to construct an ideal type of entrepreneur in a given situation. These three forms of creativity are generally enmeshed in an entrepreneur's activities, but they only produce creativity under singular conditions which do not necessarily conform to the standard description of a creative innovation process. This ideal type shows that a small business entrepreneur applies their creativity for three reasons: to resolve a problem related to regular activity, to improve parts of their activity and to address specific unusual or urgent situations. Using this analytical observation, it may be considered that a small business entrepreneur develops entrepreneurial creativity when they have neither a named research and development project, nor a pending patent application. They are actually not necessarily able to identify in themselves the link between their creativity and the concept of innovation, which, for that matter, they often uniquely define as a product innovation.

4.3. Discussion

It is therefore difficult to mark the business observed with the seal of innovation when it does not fulfill the criteria of an innovative business. However, it is clear that the entrepreneur brings new elements into play through their creativity and that this happens on a recurring basis. Their activity is continually affected by the creative approach. The creativity undertaken corresponds to an entrepreneurial path very different from the one found in the literature. We propose to disconnect the idea of creativity from that of innovation; the small business entrepreneur sees their creativity at the heart of their profession, without there actually having been a systematic intention to apply an innovation process. This activity-related creativity can make up for a poorly established routine, support and facilitate the production process, help to solve problems, and be useless or even harmful in certain cases. It deserves to be consciously assessed by the entrepreneur to determine its true usefulness. Creativity can lead to innovation in all its forms, but it does not necessarily lead to the results expected from traditional innovation processes. Entrepreneurial creativity is not innovation. On the contrary, it can be a precondition for reflection on the part of all entrepreneurs, who must be able to evaluate, classify and select which element of creativity will not be pursued and which could become a strategic focus for the business and used in a traditional innovation process. For entrepreneurs, this demarcation can uncover a source of value that deserves to be used (Figure 4.1).

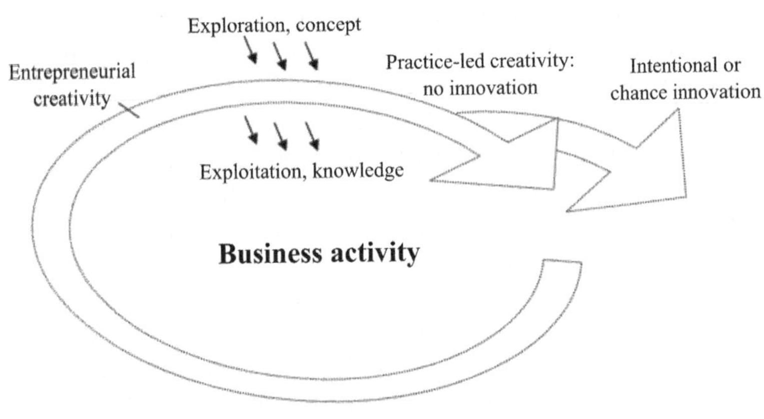

Figure 4.1. *Entrepreneurial creativity as a source of value for activity*

There are at least two aspects of entrepreneurial creativity: that which shapes activity and alters it without actually leading to an intentionally innovative approach, and that which is, or could be, a precursor to an innovative approach. The former, practice-led creativity, can emerge from the difficulty of organizing an activity, when it is difficult or impossible to apply or set up the formal processes that lead to an entrepreneur producing problem-solving ideas. This difficulty in applying established routines has its roots in the changeability of the problems encountered, a lack of training and/or education among staff, management of the time and urgent nature of tasks to be completed, executive staff reduced to a minimum, and also an estimated idea of profitability increased by the entrepreneur in the form of a perceived personal performance. The act of finding solutions or creating something new certainly contributes, as described by Maslow [MAS 43] through the hierarchization of needs, to self-actualization and the achievement of the individual entrepreneur.

Regardless of the form of creativity produced, this constitutes a creative approach unique to small business entrepreneurs as they only assess their operations in light of their own notion of a situation. If the business develops, delegation and sharing can alter this notion, and lead to greater formalization and a different creative approach, although this moves further and further from the framework of the small business. The fact that the activity is carried out with little help, and not necessarily from decision-makers, obliges the entrepreneur to introduce organizational processes only and then manage them. The emergence of strain between the creativity produced and the routines in place seems unavoidable and requires the entrepreneur to have a certain ability to multitask.

Lastly, the superfluous creativity described in some situations allows us to reflect on the hidden side of creativity. Must the entrepreneur indulge in creativity only to remake something that already exists? What is the point of inventing purely to compensate for badly executed routines? The answers are unique to each situation at each moment of the entrepreneur's activity. An economic calculation can answer this equation, taking into account, on the one hand, the costs and returns of implementing and applying a process or routine, and, on the other hand, the economics of not implementing it and the cost of the creative response. The calculation nonetheless requires the ability to foresee the emergence of a creative response, which is difficult to guarantee, especially if the entrepreneur has not made this thought central to their entrepreneurial approach. In order to carry out this calculation, on the one hand, there is a certain and determinable cost if the entrepreneur establishes a process and everything necessary to applying it, and, on the other hand, a cost that is difficult to quantify for a creative solution that is random by definition. While calculating costs is difficult, measuring profit is just as troublesome. In order to perform this calculation, the entrepreneur can mediate by introducing the idea of risk to their approach through an evaluation matrix (quantified or not) for risk, for example. Confronted with a complicated situation, the entrepreneur adopts an attitude of choice. "*They do not necessarily choose to do what they like, but choose what they would like to do*". They must then proceed to take up a creative approach tool to help with managing risk or uncertainty [MER 10]. The aim is to mark out their decision-making sphere by taking into consideration all of the aspects that could influence their decisions, whether at the strategic, tactical or operational level [ANS 65].

The development of information systems facilitates access to information for the entrepreneur, who will see a reduction in the difficulties of making choices. Outsourcing memory, for instance, improves the availability of information in all entrepreneurial situations that might arise, at the time and in the place that they may appear. A different operating mode risks appearing between those who know how to make use of new information and communication technologies, and those who, for various reasons, find themselves on the sidelines. By taking our thought further, we are nevertheless able to wonder about respect for existing formalized processes that make the production system highly efficient, but which can also become restrictive as regards the entrepreneur's creative approach. Indeed, even if this seems of little relevance when it comes to compensating for non-compliance with an established process, superfluous creativity could also lead the entrepreneur to generate serendipitous occurrences.

4.4. Conclusion

This chapter aims to open up the black box of creativity a little further and provide explanatory elements as regards its place in small businesses. The contrast of macrodata on the small business with more detailed data from observations on the ground offers us a wide-angle picture with a focus on entrepreneurial creativity. Taking creativity into account to explain entrepreneurship seems very relevant to these ideal types and allows us to suggest the deconstruction of the concepts of creativity and innovation, with creativity not necessarily causing innovation. The kinds of creativity that can appear encourage us to reflect on the issue of multitasking in small businesses. If we use the filter proposed by March [MAR 91], it is equally important for an entrepreneur to be able to examine and collect new information as to make use of it. Exploration and exploitation are essential to establishing present and future competitivity, and require an ability to multitask to balance the resulting tensions [ING 08]. The small business entrepreneur masterminds both, once again proving their abundance of creativity to themselves by handling multiple tasks simultaneously.

4.5. Bibliography

[AMA 88] AMABILE T.M., "A model of creativity and innovation in organizations", *Research in Organizational Behavior*, vol. 10, no. 1, pp. 123–167, 1988.

[ANS 65] ANSOFF H.I., *Corporate Strategy: Business Policy for Growth and Expansion*, McGraw-Hill, New York, 1965.

[ARC 91] ARCHIBUGI D., CESARATTO S., SIRILLI G., "Sources of innovative activities and industrial organization in Italy", *Research Policy*, vol. 20, no. 4, pp. 299–313, 1991.

[AUD 01] AUDRETSCH D.B., THURIK R., Linking entrepreneurship to growth (Lier entrepreneuriat et croissance), OECD Science, Technology and Industry Working Papers, 2001/2, OECD Publishing, 2001.

[BIS 11] BISSOLA R., IMPERATORI B., "Organizing individual and collective creativity: flying in the face of creativity clichés", *Creativity and Innovation Management*, vol. 20, no. 2, pp. 77–89, 2011.

[BYG 91] BYGRAVE W.D., HOFER C.W., "Theorizing about entrepreneurship", *Entrepreneurship Theory and Practice*, vol. 16, no. 2, pp. 13–22, 1991.

[CAR 97] CARRIER C., *De la créativité à l'intrapreneuriat,* Presse de l'université du Quebec, Sainte-Foy, 1997.

[CRO 10] CROSSMAN M.M., APAYDIN M., "A multi-dimensional framework of organizational innovation: a systematic review of the literature", *Journal of Management Studies*, vol. 47, no. 6, pp. 1154–1191, 2010.

[DAM 12] DAMANPOUR F., ARAVIND D., "Managerial innovation: conceptions, processes, and antecedents", *Management and Organization Review*, vol. 8, no. 2, pp. 423–454, 2012.

[DAV 04] DAVID A., "Logique, épistémologie et méthodologie en sciences de gestion: trois hypothèses revisitées", in THIÉTART R.A. (ed.), *Méthodes de recherche en management*, Dunod, Paris, 2004.

[DES 92] DESREUMAUX A., *Structures d'entreprise: analyse et gestion*, Vuibert, Paris, 1992.

[DES 03] DESCHAMPS J.P., "Innovation and leadership", in SHAVININA L. (ed.), *The International Handbook on Innovation*, Elsevier, Amsterdam, 2003.

[DES 12] DESLÉE C., GUIROD P., "La combinaison de connaissances pour innover. Le cas de la création de services d'assurances innovants pour le distributeur", *Revue française de gestion*, vol. 2, pp. 95–110, 2012.

[GAR 95] GARTNER W.B., "Aspects of organizational emergence" in BULL I., THOMAS H., WILLARD G. (eds), *Entrepreneurship: Perspectives on theory building*, Pergamon, London, 1995.

[GIR 89] GIRIN J., *L'opportunisme méthodique. La recherche action en action*, Journée d'étude AFCET, Ecole centrale de Paris, 1989.

[HAL 97] HALL H., *Organizations: Structure and Process*, Englewood Cliffs, Prentice Hall, 1972.

[HOW 98] HOWITT P., *Endogenous Growth Theory*, MIT Press, Cambridge, 1998.

[ING 08] INGRAM A.E., LEWIS M.W., ANDRIOPOULOS C. *et al.*, "Innovation tensions and organizational ambidexterity: toward a collective paradox frame", *Academy of Management Proceedings*, vol. 1, pp. 1–6, 2008.

[INP 15] INPI, Palmarès des principaux déposants de brevets publiés en 2014, Les chiffres clés 2014, 2015.

[INS 12] INSEE, Effectifs salariés par secteur et catégorie d'entreprises en 2012, Tableau de l'Economie Française, T16F153, édition 2016.

[INS 16] INSEE, Principales caractéristiques des entreprises par catégorie, Insee, Esane, Lifi, version 09 févier 2016.

[JUL 90] JULIEN P.A., "Vers une typologie multicritère des PME", *Revue internationale PME: économie et gestion de la petite et moyenne entreprise*, vol. 3, nos 3–4, pp. 411–425, 1990.

[KEU 11] KEUPP M.M., PALMIÉ M., GASSMANN O., "The strategic management of innovation: a systematic review and paths for future research", *International Journal of Management Reviews*, vol. 14, no. 4, pp. 367–390, 2011.

[LEE 93] LEE R.M., *Doing Research on Sensitive Topics*, Sage, Thousand Oaks, 1993.

[MAR 91] MARCH J.G., "Exploration and exploitation in organizational learning", *Organization Science*, vol. 2, no. 1, pp. 71–87, 1991.

[MAR 03] MARCHESNAY M., "La petite entreprise: sortir de l'ignorance", *Revue française de gestion*, vol. 144, no. 3, pp. 107–118, 2003.

[MAS 43] MASLOW A.H., "A theory of human motivation", *Psychological Review*, vol. 50, no. 4, pp. 370–396, 1943.

[MER 10] MERAD M., *Aide à la décision et expertise en gestion des risques*, Lavoisier, Paris, 2010.

[MIN 82] MINTZBERG H., *Structures et dynamique des organisations*, Editions d'Organisation, Paris, 1982.

[NYS 93] NYSTRÖM N., "Creativity and entrepreneurship", *Creativity and Innovation Management*, vol. 2, no. 4, pp. 237–242, 1993.

[PAR 15] PARMENTIER G., SZOSTAK B., "La créativité organisationnelle vue comme sous l'angle du management stratégique: proposition du concept de capacité créative d'une organisation", *Communication AIMS, XXIV Conférence Internationale de Management Stratégique*, Paris, 2015.

[SCH 09a] SCHMITT C., "Les situations entrepreneuriales: proposition d'une nouvelle grille d'analyse pour aborder le phénomène entrepreneurial", *Revue Economique et Sociale*, no. 3, pp. 11–25, 2009.

[SCH 09b] SCHMITT C., FILION L.J., "Le réel expérimenté pour explorer la TPE: apports conceptuels et méthodologiques", *Management & Avenir*, vol. 30, pp. 207–226, 2009.

[TOR 97] TORRÈS O., "Pour une approche contingente de la spécificité de la PME", *Revue internationale PME: économie et gestion de la petite et moyenne entreprise*, vol. 10, no. 2, p. 16, 1997.

[TOU 10] TOUTAIN O., Apprentissage expérientiel et métacognition dans l'éducation à l'entrepreneuriat, PhD Thesis, University of Lyon, 2010.

[TSA 14] TSAI K.C., "Creativity is the spirit of entrepreneurship", *International Journal of Business and Social Sciences*, vol. 2, no. 1, pp. 106–115, 2014.

[VAN 86] VAN DE VEN A.H., "Central problems in the management of innovation", *Management Science*, vol. 32, no. 5, pp. 590–607, 1986.

[VER 03] VERSTRAETE T., *Proposition d'un cadre théorique pour la recherche en entrepreneuriat*, Editions de l'Adreg, Paris, 2003.

[VER 05] VERSTRAETE T., FAYOLLE A., "Paradigmes et entrepreneuriat", *Revue de l'Entrepreneuriat*, vol. 4, no. 1, pp. 33–52, 2005.

[WOO 93] WOODMAN R.W., SAWYER J.E., GRIFFIN R.W., "Toward a theory of organizational creativity", *Academy of Management Review*, vol. 18, no. 2, pp. 293–321, 1993.

Other sources

[BPI 15] BPI, Rapport annuel sur l'évolution des PME, Observatoire des PME, available at: http://library.bpifrance-lelab.fr/PME15, 2015.

[CIR 13] CIRCULAIRE MINISTÉRIELLE EN-ESR, Circulaire sur la création du statut national d'Etudiant-Entrepreneur, Ministère de l'Education Nationale, de l'Enseignement supérieur et de la Recherche, available at: www.enseignementsup-recherche.gouv.fr/cid74716/pepite-constitution-de-poles-etudiants-pour-l-innovation-le-transfert-et-l-entrepreneuriat.html, 18 March 2013, 2013.

[EUR 05] EUROSTAT, Manuel d'Oslo: principes directeurs pour le recueil et l'interprétation des données sur l'innovation, OECD Publishing, 2005.

[INS 14] INSEE, Stock d'entreprise et entreprises créées en 2014, France et Insee Première, no. 1521, October 2014.

5

From Ideation to Product Launch

Innovation is determined by the ability to identify connections, discover opportunities and make a profit from them, by opening up new opportunities (in which technology often plays a crucial role) or by inventing new methods that are able to satisfy the needs of existing markets. An interesting example of this type of improvement is given, for the record, in the context of frugal innovation intended for developing countries, by a pair of glasses with lenses that can be adjusted by the patient according to their specific needs [TID 06], without needing to resort to an optician of whom there could be an acute shortage (the manufacturing technique is designed to maximize the quality/price ratio). The remainder of the chapter is dedicated to a high-tech business environment in the same area as the previous example, but in the field of researching technological excellence. The ophthalmic industry is characterized by the increasing complexity of its vision correction products which has been brought about by ever more specialized knowledge at all stages of design and manufacturing. This strong trend leads to concerns about insourcing/outsourcing strategies for activities requiring extensive knowledge, which are more demanding when the field is highly competitive. The issue with these strategies therefore hinges on the approach to outside collaboration in innovation management, an approach in which the contractualization of intellectual property rights is shown to be just as essential as galvanizing colleagues (within the company) using outside knowledge. In fact, as regards this input, recent works have shown that geographically localized access to knowledge directly influences a firm's innovation capabilities and levels of R&D investment, as well as the amount

Chapter written by Maggy PERRIER and Audrey DEPEIGE.

and type of knowledge acquired from outside, in relation to the value of social capital [LAU 12]. The aim of this chapter is therefore to analyze this issue by illustrating it through the perspective of the French company Essilor.

In this regard, it seems crucial for the organization to explain both the opportunities and the challenges that engineers face when they endeavor to gather and combine knowledge [MUR 07]. Furthermore, the company will probably benefit from the inclusion of socio-ecological factors in its efforts to understand and systematically evaluate its competitive environment. Such efforts boost innovation performance through the generation and implementation of strategic foresight [PET 09], but this raises the question of the acquisition of the input in question and its integration by engineers in the business's innovation procedure; the process of interweaving these different areas of knowledge within a successful innovation unfolds in highly uncertain conditions and managing the innovation consists of transforming these uncertainties into new knowledge [TID 06].

This analysis will be carried out in three steps, all illustrated with the example of Essilor. The first step will set up the framework for collaborative innovation, starting with the innovation mechanisms activated by internal actors and going on to show how they are enhanced by outside input. The second part will address the organizational challenges by going into greater depth on aspects within the business. The management of intellectual property rights, both for internal activity and external collaborative activity, will be the subject of the third part. The conclusion will highlight the challenges of managing innovation in an international business and the solutions provided by the intellectual property strategy that is considered the strong arm of the collaborative innovation strategy in presiding over knowledge exchanges and allowing that knowledge to be transformed into value. It therefore remains aware that the success of collaborative innovation is highly dependent on the organization and quality of the training provided.

5.1. The quest for continuous renewal: collaborative innovation within business strategy

The pressure of intense competition has pushed organizations into adopting systems that boost knowledge transfer and retention, while also creating value by producing knowledge. From a theoretical perspective of knowledge-based innovation, innovation is defined as a new and useful product of knowledge that can be reproduced [QUI 11]. The deployment, integration and usage of knowledge resources intercede between intangible, intra-organizational assets and results in terms of the growth and performance of innovation [MAU 11]. While such systems have brought about a resurgence of interest from researchers and professionals, research led by Vasudeva *et al.* [VAS 13] suggests that the competitive and innovative capabilities of a business can be improved by partnerships with other strong actors owning significant social capital. This being the case, the businesses can create value and growth, as well as new technological knowledge and legitimacy by gaining access to new knowledge resources. However, previous research has shown that a random search for resources in the business environment cannot provide convincing results [BUT 06]. In fact, the challenging global landscape requires companies to adapt to change in order to benefit from emerging opportunities on the global market. In this regard, it becomes particularly important to establish new collaborations and new partnerships.

Ketchen *et al.* have argued that collaborative innovation provides large organizations with opportunities to meet new challenges by exploiting resources such as networks and learning capabilities, or by seizing opportunities in order to share, create and develop the potential for knowledge and innovation [KET 07]. In this perspective, it seems that organizations would profit from seizing opportunities (particularly from technological advances) to develop new partnerships, which would help them adapt to environmental changes. Specifically, experience on the ground has shown that a rapid pace of development in technologies and markets is not (or is no longer) always enough to guarantee the development of competitive products using only internal resources[1]. As suggested by the recent growth in initializing external collaborations, firms strive to explore combinations of expertise and flexibility by collaborating with other firms in innovation networks [GUR 09]. Such scenarios mainly emerge from increasing the complexity of innovation procedures and technologies which are not necessarily accessible internally. Illustrating this point of view, Essilor takes advantage of long-standing (or more recent) partnerships with local universities which allows its internal innovators to further develop their

1 The dynamics of internal resources, in terms of implementing a knowledge strategy as a method of investigating as much as utilizing, struggle to follow the dynamics of the markets and cutting-edge technology.

knowledge, capabilities and networks by simultaneously optimizing the investments and operational costs of developing a product. The development of the Prevencia® product illustrates this point well, as Essilor had led an earlier collaboration with the Institute of Vision (*Institut de la vision* in French), thereby becoming the first lens manufacturer to launch a clear lens that offered eye protection by filtering UV rays and blue light from screens with low reverse reflection. Moreover, in order to transform the idea into an innovative product, the company follows a detailed process with pre-determined stages as the collection and identification of relevant knowledge to acquire from external environments, which will be re-used to resolve internal business challenges, as well as to contribute to improving current performance. A process of discovering and formalizing ideas, with the aim of transforming ideas into practical inventions, has thus been rolled out (see Figure 5.1).

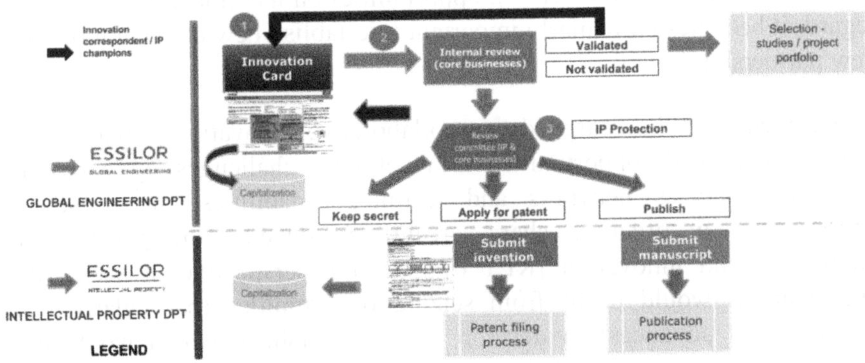

Figure 5.1. *Diagram of the innovation procedure implemented by Essilor (source: Essilor)*

First of all, it is clear that the transformation of a creative idea into a marketable innovation requires significant work so that the final product becomes a best-seller. To facilitate the work of describing and identifying new ideas early on in the proceedings, Essilor developed an “innovation file” that briefly describes and graphically illustrates the idea. The aim of the innovation file is to capture the “inventive idea” that could become a product or design tool by building on the vision of innovation defined by Essilor. As such, the author of the invention is encouraged to present his/her idea through the innovation file and, in order to do so, he/she is continually guided by “innovation correspondents” or “intellectual property champions” (1). This phase illustrates work autonomy well, in the sense that each inventor is encouraged to develop ideas throughout the innovation process. While the idea generation phase focuses on dynamic and inter-team methods of producing a range of ideas in brainstorming sessions, the later stages are dedicated to evaluating ideas, integrating them and building innovation by means of the innovation file (2). During these stages, an in-depth analysis is conducted by business, and the opportunities for creating intellectual property rights are identified. As engineers may reason analogously and focus on solutions very early in the design process, specific creative problem-solving methods are implemented in addition to idea generation methods to draw creative ideas from existing solutions (typically, engineers rely on "design thinking" methods or "concept-knowledge driven" tools). The team of engineers in charge of the project is therefore asked to assess the feasibility of the idea and outline the possibilities for manufacturing and distributing the product. The ideas that successfully emerge from this analysis then follow the IP procedure in accordance with three possible directions (3), with the process transversal to the different organizations and being supported by the Executive Committee.

Concurrently with existing processes, the managers are evaluated as much on their ability to support colleagues in their generation of new knowledge as on their ability to exploit knowledge already available within the business [NIC 04]. This premise is also supported by other works, which highlight the fact that, even if an organization's capability to acquire and store knowledge is absolutely essential, it does not necessarily lead to innovative outcomes, unless knowledge has been exploited, processed and recombined in such a way as to shape the development of innovative ideas and increase the probability of inventions reaching a stage of maturity [JOS 10]. In such cases, for Provost and Fawcett, organizations should split their business problems into smaller elements to be able to assess the prospects and estimate expected value out of the innovation [PRO 13]. However, the history of the firm shows that conditions for cumulative innovation are established at multiple levels [MUR 07]: the institutional level (regulatory institutions), the field level (market), the organizational level (firm) and the community level (society). At the institutional level, Essilor collaborates with, for example, the University of Shanghai to pool their available skills and means to investigate and design dedicated solutions. At the field level, there is the social innovation program EyeMitra, which consists of an innovative business model which has been rolled out in rural India. The initiatives and means established at the organizational level are varied. They include awareness channels, research departments, transversal technical communities and national innovation laboratories, as well as the Nikon-Essilor International Joint Research Center, particularly thanks to significant investment from R&D turnover while other internal structures, such as Corporate Social Responsibility (CSR), simply strengthen the numerous controls in technical expertise and collaborative problem solving to develop cutting-edge solutions regarding eyesight. By focusing on business innovations with a social impact, Essilor has recently launched an open innovation contest named "See Change" aiming to crowdsource expertise among universities, scientists, developers, engineers and start-ups to develop new solutions for 2.5 billion people who do not yet have access to corrective lenses.

In correlation, previous research tends to show that the exchange of knowledge and the development of interdisciplinary interactions between stakeholders contribute to the generation of a significant quantity of results both in terms of innovation and organizational performance [HAR 97, OBS 05]. It follows that this strategic approach can boost innovation in several aspects at the same time. In this sense, collaboration between firms, government agencies, research institutions and investors represents a catalyst for exchanging new forms of knowledge and developing high-level innovation, while also creating value for the innovation's co-creators. In that respect, networks facilitate the circulation of flows of information across organizational boundaries; research highlights the fact that organizations in networks tend to increase the ability of firms to survive, while also reducing competition [SOL 10]. This is shown at Essilor in the growth of partnership networks outside the firm (as illustrated above), with an acquisition policy that will enhance networks for distribution, production and trade. This allows the firm to offer innovative products and services adapted to the needs of different groups of consumers and the expectations of local markets covering a very large part of the vision correction and protection firm. Synergies between traditional and newly acquired industrial or business units are systematically encouraged (within the boundaries of respecting competition) in order to nurture innovation while optimizing the necessary resources. Taking this further, it seems that interactions between different departments allow a breeding ground to be created in a specific service, which can itself go on to feed local knowledge and provoke the emergence of new knowledge. Such mechanisms indicate that the relevance of the knowledge possessed by some departments will probably affect the quantity and destination of the knowledge distributed. Alcater and Zhao [ALC 10] therefore defend the idea that close links between industrial locations allows firms to greatly increase knowledge to gain competitive advantages – by pooling "the advantages of a location, such as proximity to research universities, favorable government policies, and an abundance of human capital" – while innovations from firms co-located with their competitors tend to quickly become insourced.

Moreover, the results of previous research reveal that change within an environment allowing access to new resources and opportunities is a key factor for innovation in firms [HEA 06], which is consistent with the results of Cooms *et al.* [COO 09] and Hmielski and Corbett [HMI 08], who believe that the geographic background of the organization dramatically increases performance in dynamic environments, either because of local and international research strategies or steady internal research. Similarly, academic research indicates that the firm's level of international experience adversely reduced the outcomes of a study on expanding internationalization, explaining that "the lower a firm's experiential learning, the more significant the effects of the prior international base and the knowledge and skills acquired through key partners" [BRU 10]. To complete the point on the positive impact of networks, Essilor simultaneously maintains a balance between local services and operations, and "corporate" teams whose sphere of activity and international presence significantly contribute to the market capitalization and distribution of knowledge, expertise and good practices within the firm.

More specifically, a number of studies have underlined the beneficial influence of the firm's external environment, as well as its individual capabilities, on the efficiency of the two types of innovation (exploration and exploitation) and on fostering learning from these perspectives [KAN 07]. Nonetheless, the performance of the organization is dependent on effectively establishing a balance between the two kinds of dynamics, namely organizational sustainability and learning. As well as its R&D investments and patents, the firm's innovation strategy can include value creation and access to specific technologies by means of mergers/acquisitions, and thereby relying on external knowledge; however, it is unlikely that "innovation strategy can be charged using a simple make vs. buy logic" [ARO 13]. The example of Essilor's entry into the "readers" market (lenses correcting near-sightedness and based on fixed prescription) illustrates the strategic choice of developing this activity acquired through the acquisition of the company FGX by including a selection of innovations from its providers. Outsourcing is therefore a quick and optimal way of offering clients the latest innovations in addition to R&D activities.

5.2. An internal environment supporting innovation

In light of previous examples, it is clear that a key aspect of innovation lies in the success of commercial applications and the constant and systematic submission of new ideas to the market [SHA 03]. Kuratko *et al.* [KUR 13] identified five aspects of the firm's internal environment: top management support, work autonomy, rewards, time availability and organizational boundaries. This research highlights the role of diagnostic tools in evaluating the parameters that contribute to creating a team environment conducive to innovation (audits on maturity and innovation) that allows the maturity of the business to be assessed in relation to these five factors and emphasize the areas of the firm that need to improve in supporting development. The authors complement this view by suggesting that it is the responsibility of the managers to create a work environment that is "highly conducive to innovation and entrepreneurial behavior" (p. 39). Illustrating this point, it seems that new opportunities could emerge in the form of new products, new services, new instruments/systems or incremental improvements in production methods or existing products. Drucker [DRU 85] noted that diversity, expressed through a range of complementary procedures (Essilor's internal creativity and innovation processes), industrial structures (Essilor's partnerships with suppliers to develop innovations in manufacturing processes, for example), market demographics (competitive intelligence efforts), changes in reference frameworks (deploying the "design thinking" approach) and knowledge (deploying the C-K method) are the sources of new opportunities leading to innovation. By bringing all of this together, we can see that collaborative relationships based on and developed under these values are likely to lead to the development of innovations that strengthen or extend capital in all aspects (social, economic and environmental) of a firm, whose staff is the cornerstone. Specifically, Kuratko *et al.* [KUR 13] note that collaborators play a key role in creating environments favorable to innovation, such as "stimulating practices, devoting time to unproven approaches, persevering in the face of organizational resistance (…)". These authors insist that enterprises should devote significant efforts to finding ways and means to sustain an innovative working environment despite the firm's environment constantly undergoing change:

1) Managerial support: at Essilor, such support organizes continuous efforts to foster an innovative culture through innovation contests or measuring the outcomes of innovation with "Balance Score Cards".

2) Work autonomy: strengthening the bonding of teams across organizational boundaries and facilitating the transformation of creative ideas into innovations with market launch potential. Essilor has recently introduced "IP champions", whose role is to interface the promotion of innovative solutions between business/operational teams (defining the solution) and Intellectual Property teams (assessing the technical value of the idea and its patentability).

3) Rewards: both tangible (patent monetary-based rewards) and intangible rewards (annual inventor award ceremonies) are covered in Essilor programs that aim to encourage creativity and innovation within the company.

4) Time availability: management of Essilor's portfolio of projects takes into consideration the readiness of solutions and a process of selection and prioritization, which allows some resources to be dedicated to exploratory and innovative projects.

5) Organizational boundaries: in order to complete the innovation card system, a pipeline process combined with a "stage gate" approach has been introduced to maximize the chances of gathering all creative ideas of value in different departments of the company, regardless of the creator, and of being able to then provide means and support engineers in testing ideas, from the proof of concept to the launch on to the market. It is a matter of dividing projects according to their maturity, and then of pushing them through the process ("gates" passages) with respect to their stage of development, with the final selection of projects being made on the basis of the value brought to the firm. This pipeline process particularly allows costs to be optimized and, simultaneously, the success of the selected projects to be guaranteed.

In line with these five dimensions, Saint-Onge [SAI 96] argues that the cross of human capital, internal resources and external structures forms the basis for value creation. In other words, successful companies are able to generate value by integrating both tacit and explicit knowledge, as well as expertise and awareness of external realities. Such an approach requires existing knowledge (implicit or explicit) to be transformed into broader knowledge structures which build up systemic knowledge [NON 95]. Schulz [SCH 03] noted that the relevance of new knowledge determines the extent to which it can be connected to existing knowledge. According to this author, external knowledge has more potential to connect local knowledge if its implication for local knowledge is higher and easier to derive. As an illustration, engineering teams are led towards exploring three different types of knowledge bases which provide an essential background for the ways innovative ideas emerge within the organization; technological knowledge (knowledge of engineering and technical aspects of products), market knowledge (marketing and sales channels) and knowledge of the competitive landscape (including suppliers, government agencies and competitors). Concretely, engineering teams enrich their ideas with the support of transversal functions such as "techno plans" and "competitive intelligence", incorporating insights from upstream functions such as Marketing, thereby integrating benchmark data, economic and competitive intelligence, as well as the latest market developments to accommodate the needs of lenses holders. Furthermore, consumer-oriented "road maps" have been developed within R&D, aiming to leverage the knowledge gained at the company level while also assessing its potential – Should it be in terms of products or service development – by targeting the needs of the end customers. Roles emerging from the roadmap are operated transversely, together with multidisciplinary teams. In other words, technical and scientific teams are able to evolve and enrich their knowledge bases by combining new knowledge and prior related knowledge. It seems that a combination of internal resources, external networks and knowledge acquisition strengthens companies' ability to create value and new knowledge, and therefore to innovate. The ability to build long-term value is deemed essential to firms, and, more specifically, R&D-based organizations operating in knowledge-intensive activities.

Previous research has claimed that combining multilevel approaches, complementary tactics and synergistic learning contributes to reinforcing organizational ambidexterity [AND 09b]. Similarly, for Anderson *et al.* [AND 09a], it is the capability of firms to be entrepreneurial by generating and acting on strategic knowledge that contributes to advancing organizational learning, which, at the firm level, is illustrated by the encouragement of engineers of all backgrounds to design and prototype their innovative solutions. Moreover, Holmqvist [HOL 04] describes processes related to organizational learning as involving openness to new ideas, which in turn requires the firms to enter an exploratory process, such as experimentation and trialing, rather than continuing with the routine and repetitive use of existing knowledge. Colombelli *et al.* [COL 12] support this view, arguing that firms with strong knowledge tend to predominantly explore knowledge, while displaying tendencies to organize search properties more often associated with knowledge exploitation strategies. Both knowledge dynamics are called upon to solve different kinds of problems (single-objective vs. multi-objective) with various stakeholders involved to obtain results [CRE 13]. Learning benefits both exploratory and exploitation knowledge strategies [SIR 12]. According to these authors, "the process of strategic learning, through its intra-organizational elements, enables the dissemination, interpretation, and implementation of strategic knowledge, and enables firms to capitalize on the benefits of both exploration and exploitation strategies". Murray and O'Mahony [MUR 07] indicated a lack of research in this area, emphasizing the need to understand how organizations exploit the opportunities and manage the challenges of knowledge gathered within firms as well as across organizational boundaries. Essilor combines the "exploitation" innovation strategy with incremental innovations such as Varilux® and Crizal® where the company has developed strong development processes that rely on internal skills, and the "exploration" innovation strategy in little developed sectors such as the area of "next generation consumers", for which the traditional approach does not always provide responses suited to the requirements of the 2.5 billion people worldwide who do not yet have access to vision correction.

5.3. Managing invention through innovation: building a strong intellectual property management process to maintain a competitive advantage

Value creation in firms is increasingly conceptualized through a prism of ecologies [HEA 06], which illustrates a shift in both our understanding and analysis of innovation phenomena. As regards the value chain, relationships and value creation processes are increasingly analyzed as emerging from an ecosystem, with a focus on value networks and value co-creation [BEN 13]. An argument for this paradigm change is that the ecological view puts an emphasis on the dynamics between actors, which is supported by studies pointing out the shift of market interactions from transactions to relationships [GUR 09]. In addition, sustainable development literature constitutes the relevant framework for analyzing innovation and business expansion, pointing out the role of new regulations, which push for the development of ideas and innovations leading to business growth; this is the "regulatory push and vision pull" model [PRE 07]. From this perspective, the process of reiteration and the feedback loop are an important element of value creation ecologies, contributing to establishing multidirectional networks as inherent to value creation [HEA 06]. Nevertheless, the ecosystem view requires incorporating the risk of conflict that may emerge from tensions between firms [HEA 06], as the value creation ecosystem suggests a much broader impact on the capture and creation of value, which is built holistically rather than individually and which extends far beyond partnerships between firms or between individuals that aim to increase an actor's capabilities over the longer term. The formalization and contractualization of patent terms is a key aspect of the external cooperation developed by Essilor, for example in the field of equipment where, frequently, competitors share the same suppliers for specific technologies; these can be, for instance, innovative production processes owned and developed by Essilor and technical expertise provided by the supplier, including acquisitions. In this context, internal links and partnerships are key elements for extracting the maximum value from innovations [ALC 10]. The business is particularly reliant on its ecosystem in the field of equipment, as skills in that area are no longer developed internally, and it works closely with suppliers, including on the development of innovations by suppliers.

The theory of intellectual capital, moreover, suggests that there are cases where processes and technologies must be removed from competing partners [SOL 10] to avoid the risk of losing a competitive advantage. Alcater and Zhao [ALC 10] state that "as long as the innovating firm can integrate locally developed knowledge more efficiently than potential imitators, it can take advantage of the low cost in weak IPR countries and still gain from innovation". However, companies do not always have the financial means to invest in new forms of technologies, systems and knowledge, which restricts their possibilities for growth. Moreover, such international configurations require formalities to develop and support the organizational structures that enable them to maintain strict control of their innovations, sometimes engaging in learning or patent races. Colombelli *et al.* [COL 12] recalled that technology creation is often mandated by patent application and that it follows a series of outputs, appearing as a means of financing knowledge creation stages. Salerno *et al.* [SAL 15] specify that, although firms tend to use a linear innovation process internally, the projects and related patents emerging from an innovation development usually involve different companies, with each stakeholder having different roles, different responsibilities and different contributions. The firms involved may operate on different markets, for example, serving different clients, making use of different technologies and benefiting from different regulatory conditions, which is the case with Satisloh, an equipment company proposing machining and process solutions for the entire ophthalmic industry, which was acquired by Essilor in 2008, and already serves market competitors. Besides influencing the nature, scope and extent of collaboration between firms, such differences are likely to affect the outcomes of innovation process, that is, patenting and commercialization of the innovative product. As asked by Moser [MOS 13], is there an optimal intellectual property rights system specific to encouraging innovation? The intellectual property strategy can, for instance, be shared between the development of internal innovations and the imitation of innovations made externally by suppliers, in which case the property is shared within the business's ecosystem. In fact, ideas for a new business can actually emerge beyond yet also within the boundaries of the organization. Firms normally use both internal and external sources in order to access information, technologies, innovation, business practices and/or networking with other companies to be able to increase growth and profitability [NAR 09]. In such cases, firms may need new appliances, people and knowledge [HUS 04a]. Previous research has highlighted that a "successful innovation strategy requires careful alignment both between internal and external knowledge sourcing" [ARO 13], suggesting that extensive technical knowledge is not enough for a firm to overcome knowledge gaps. This indicates that close collaboration, new forms of partnership and knowledge networking can result in inter-organizational learning and dissemination of new, valuable information. Powell *et al.* [POW 05] highlighted the importance of participating in such networks due to the crucial growth factors they provide, such as access to new forms of knowledge, the reliability of information and adaptiveness to change. This also affects knowledge acquisition strategies [ALC 10], where a company may file multiple patent applications for a given innovation (for example, the Varilux® S Series® products are protected by around 10 patents). The company will particularly rely on internal knowledge linked to the technological innovation products, while also researching new services that create value for its consumers in order to differentiate itself from its competitors.

In this way, and while tangible resources are either easily accessible or subject to imitation by competitors, intangible resources, such as knowledge, constitute an invaluable capital asset that leads the firm to intra-organizational achievements that can only later be transformed into market successes. In that respect, confidentiality agreements between Essilor and its development partners are fundamental to ensuring the success of their relationships, regardless of how new the partnership may be. Husted and Vintergaard [HUS 04b] drew attention to the potential of turning intangible knowledge assets (ideas) into commercially exploitable intangible assets (inventions). The strength of intangible resources lies in coordinating existing tangible resources to generate innovation and competitive advantages, thereby securing continuous business development. The value appropriation complexities introduced when engaging in collaborative technological activities may exceed than their value-enhancing potential [BEL 10]. Going further, it seems that the share of innovation that companies choose to patent varies across sectors. As an alternative to patenting, industrial secrecy may still be used as a way to protect innovative products or production improvement mechanisms, while scientific publications may cover innovative technology.

5.4. Conclusion

Over the last decade, academic literature has regularly insisted on the role of dynamic capabilities in maintaining the innovation performance of firms. This chapter provides numerous insights about the methods and challenges of managing innovation within an international firm. We have specifically described the different ways in which Essilor defines and supports its innovation processes (internally and/or externally), by emphasizing the role of the internal work environment to support innovation development. Moreover, the duality of business interests in value creation and appropriation appears in the development of the company's intellectual property strategy, by distinguishing the flows of value creation (commercializing innovation) and value appropriation (whether or not to patent the invention). The chapter has also mentioned how processes involving external partners differ in terms of scope and legal restrictions, and how they have implications beyond the linear mechanism implied by the "from idea to innovation" sequence. In particular, by sharing elements of its knowledge with other actors, Essilor aims and conspires to increase value creation at the level of its own ecosystem, while simultaneously reducing costs – by sharing resources – and benefiting from capturing knowledge objects, especially through interactions with its suppliers. Knowledge therefore appears to be a key factor in identifying new sources of internal and external growth, as well as the possibilities for innovation that they represent. In company networks, in particular, it is suggested that knowledge is not only distributed within the network, but is also combined with other sources of knowledge (exclusive or not) and built on the basis of a dynamic and multidimensional process.

5.5. Bibliography

[ALC 10] ALCATER J., ZHAO M., "Local R&D strategies and multi-location firms: the role of internal linkages", *Management Science*, vol. 58, no. 4, pp. 734–753, 2010.

[AND 09a] ANDERSON B.S., COVIN J.G., SLEVIN D.P., "Understanding the relationship between entrepreneurial orientation and strategic learning capability: an empirical investigation", *Strategic Entrepreneurship Journal*, vol. 3, no. 3, pp. 218–240, 2009.

[AND 09b] ANDRIOPOULOS C., LEWIS M.W., "Exploitation-exploration tensions and organizational ambidexterity: managing paradoxes of innovation", *Organization Science*, vol. 20, no. 4, pp. 696–717, 2009.

[ARO 13] ARORA A., BELENZON S., RIOS L.A., "Make, buy, organize: the interplay between research, external knowledge, and firm structure", *Strategic Management Journal*, vol. 35, no. 3, pp. 317–337, 2013.

[BEL 10] BELDERBOS R., FAEMS D.L., "Technological activities and their impact on the financial performance of the firm: exploitation and exploration within and between firms", *Journal of Product Innovation Management*, vol. 27, no. 6, pp. 869–882, 2010.

[BEN 13] BEN LETAIFA S., GRATACAP A., ISCKIA T., *Understanding Business Ecosystems: How Firms Succeed in the World of Convergence*, De Boeck, Brussels, 2013.

[BRU 10] BRUNEEL J., YLI-RENKO H., CLARYSSE B., "Learning from experience and learning from others: how congenital and interorganizational learning substitute for experiential learning in young firm internationalization", *Strategic Entrepreneurship Journal*, vol. 4, no. 2, pp. 164–182, 2010.

[BUT 06] BUTEL L., WATKINS A., "Clusters of entrepreneurs: the application of ant colony optimisation modelling", *Journal of Modelling in Management*, vol. 1, no. 3, pp. 255–269, 2006.

[COL 12] COLOMBELLI A., KRAFT J., QUATRARO F., "High-growth firms and technological knowledge: do gazelles follow exploration or exploitation strategies?", *Industrial and Corporate Change*, vol. 23, no. 1, pp. 261–291, 2012.

[COO 09] COOMS J.E., DEEDS D.L., DUANE IRELAND R., "Placing the choice between exploration and exploitation in context: a study of geography and new product development", *Strategic Entrepreneurship Journal*, vol. 3, no. 3, pp. 261–279, 2009.

[CRE 13] CREPINSEK M., LIU S.H., MERNIK M., "Exploration and exploitation in evolutionary algorithms: a survey", *ACM Computing Surveys (CSUR)*, vol. 45, no. 3, p. 35, 2013.

[DRU 85] DRUCKER P., *Innovation and Entrepreneurship: Practice and Principles*, Harper and Row, New York, 1985.

[GUR 09] GURAU C., "From information management to knowledge management", in KHOSROW-POUR M., *Encyclopedia of Information Science and Technology*, 2nd ed., Information Science Reference, 2009.

[HAR 97] HARGADON A., SUTTON R.I., "Technology brokering and innovation in a product development firm", *Administrative Science Quarterly*, vol. 42, no. 4, pp. 716–749, 1997.

[HEA 06] HEARN G., PACE C., "Value-creating ecologies: understanding next generation business systems", *Foresight*, vol. 8, no. 1, pp. 55–65, 2006.

[HMI 08] HMIELSKI K.M., CORBETT A.C., "The contrasting interaction effects of improvisational behavior with entrepreneurial self-efficacy on new venture performance and entrepreneur work satisfaction", *Journal of Business Venturing*, vol. 23, no. 4, pp. 482–496, 2008.

[HOL 04] HOLMQVIST M., "Experiential learning processes of exploitation and exploration within and between organizations: an empirical study of product development", *Organization Science*, vol. 15, no. 1, pp. 70–81, 2004.

[HUS 04a] HUSSI T., "Reconfiguring knowledge management-combining intellectual capital, intangible assets and knowledge creation", *Journal of Knowledge Management*, vol. 8, no. 2, pp. 36–52, 2004.

[HUS 04b] HUSTED K., VINTERGAARD C., "Stimulating innovation through corporate venture bases", *Journal of World Business*, vol. 39, no. 3, pp. 296–306, 2004.

[JOS 10] JOSHI K.D., CHI L., DATTA A. *et al.*, "Changing the competitive landscape: continuous innovation through IT-enabled knowledge capabilities", *Information Systems Research*, vol. 21, no. 3, pp. 472–495, 2010.

[KAN 07] KANE G.C., ALAVI M., "Information technology and organizational learning: an investigation of exploration and exploitation processes", *Organization Science*, vol. 18, no. 5, pp. 796–812, 2007.

[KET 07] KETCHEN D.J., IRELAND R.D., SNOW C.C., "Strategic entrepreneurship, collaborative innovation, and wealth creation", *Strategic Entrepreneurship Journal*, vol. 1, nos. 3–4, pp. 371–385, 2007.

[KUR 13] KURATKO D.F., HORNSBY J.S., COVIN J.G., "Diagnosing a firm's internal environment for corporate entrepreneurship", *Business Horizons*, vol. 57, no. 1, pp. 37–47, 2013.

[LAU 12] LAURSEN K., MASCARIELI F., PRENCIPE A., "Regions matter: how localized social capital afects innovation and external knowledge acquisition", *Organization Science*, vol. 23, no. 1, pp. 177–193, 2012.

[MAU 11] MAURER I., BARTSCH V., EBERS M., "The value of intra-organizational social capital: how it fosters knowledge transfer, innovation performance, and growth", *Organization Studies*, vol. 32, no. 2, pp. 157–185, 2011.

[MOS 13] MOSER P., "Patents and innovation: evidence from economic history", *The Journal of Economic Perspectives*, vol. 27, no. 1, pp. 23–44, 2013.

[MUR 07] MURRAY F., O'MAHONY S., "Exploring the foundations of cumulative innovation: implications for organization science", *Organization Science*, vol. 18, no. 6, pp. 1006–1021, 2007.

[NAR 09] NARAYANAN T.R., "Academia–industry partnership: an impetus for strengthening teaching and research in higher education institutions", *Current Science*, vol. 96, no. 3, pp. 343–346, 2009.

[NIC 04] NICKERSON J.A., ZENGER T., "A knowledge-based theory of the firm – The problem-solving perspective", *Organization Science*, vol. 15, no. 6, pp. 617–632, 2004.

[NON 95] NONAKA I., TAKEUCHI H., *The Knowledge-Creating Company: How Japanese Companies Create the Dynamics of Innovation*, Oxford University Press, 1995.

[OBS 05] OBSTFELD D., "Social networks, the tertius iungens orientation, and involvement in innovation", *Administrative Science Quarterly*, vol. 50, no. 1, pp. 100–130, 2005.

[PET 09] PETRINI M., POZZEBON M., "Managing sustainability with the suport of business intelligence: integrating socio-environmental indicators and organisational context", *The Journal of Strategic Information*, vol. 18, no. 4, pp. 178–191, 2009.

[POW 05] POWELL W.W., WHITE D.R., KOPUT K.W. *et al.*, "Network dynamics and field evolution: the growth of interorganizational collaboration in the life sciences", *American Journal of Sociology*, vol. 110, no. 4, pp. 1132–1205, 2005.

[PRE 07] PREUSS L., "Contribution of purchasing and supply management to ecological innovation", *International Journal of Innovation Management*, vol. 11, no. 4, pp. 515–537, 2007.

[PRO 13] PROVOST F., FAWCETT T., *Data Science for Business: What You Need to Know about Data Mining and Data-analytic Thinking*, O'Reilly Media Inc., Sebastopol, 2013.

[QUI 11] QUINTANE E., MITCH CASSELMAN R., SEBASTIAN REICHE B. *et al.*, "Innovation as a knowledge-based outcome", *Journal of Knowledge Management*, vol. 15, no. 6, pp. 928–947, 2011.

[SAI 96] SAINT-ONGE H., "Tacit knowledge the key to the strategic alignment of intellectual capital", *Planning Review*, vol. 24, no. 2, pp. 10–16, 1996.

[SAL 15] SALERNO M.S., DE VASCONCELOS GOMES L.A., DA SILVA D.O. *et al.*, "Innovation processes: which process for which project", *Technovation*, vol. 35, pp. 59–70, 2015.

[SCH 03] SCHULZ M., "Pathways of relevance: exploring inflows of knowledge into subunits of multinational corporations", *Organization Science*, vol. 14, no. 4, pp. 440–459, 2003.

[SHA 03] SHANE S.A., *A General Theory of Entrepreneurship: The Individual-Opportunity Nexus*, Edward Elgar Publishing, 2003.

[SIR 12] SIREN C.A., KOHTAMAKI M., KUCKERTZ A., "Exploration and exploitation strategies, profit performance, and the mediating role of strategic learning: escaping the exploitation trap", *Strategic Entrepreneurship Journal*, vol. 6, no. 1, pp. 18–41, 2012.

[SOL 10] SOLITANDER M., TIDSTROM A., "Competitive flows of intellectual capital in value creating networks", *Journal of Intellectual Capital*, vol. 11, no. 1, pp. 23–38, 2010.

[TID 06] TIDD J., BESSANT J., PAVITT K., *Management de l'innovation ; intégration du changement technologique, commercial et organisationnel*, De Boeck, Brussels, 2006.

[VAS 13] VASUDEVA G., SPENCER J., TEEGEN H., "Bringing the institutional context back in: a cross-national comparison of alliance partner selection and knowledge acquisition", *Organization Science*, vol. 24, no. 2, pp. 319–338, 2013.

6

The Patent: A "Swiss Army Knife" for Invention and Innovation

The patent system was created over two centuries ago to speed up technical progress, thanks to a clear acknowledgment of creators and good knowledge sharing practices. From the days of the quill to the Internet, the world has undergone significant changes! However, the founding principle behind patents is more relevant than ever.

Moreover, with the Taylor system having been applied to activities, patents have largely become a matter for specialists. The majority of innovators have little knowledge of patents.

However, for several months, thanks to a graphic representation of patent claims called the "means tree", all innovators are able to understand the patents in their technical field. This opens up new opportunities which this chapter will begin to explore.

After having defined some key terms, this chapter will present a general picture of the relationships that exist between innovation, inventions and patents (the second part of the chapter).

The third part will present the means tree and its applications, which allow all stages of the innovation process to be stimulated.

Finally, the fourth part will explain how the patent system can serve as a model for knowledge management dedicated to innovation.

Chapter written by Yann de KERMADEC.

6.1. Invention, innovation and intellectual property: some references

6.1.1. *Definitions*

"I will not discuss a word as long as we have not defined it" (Blaise Pascal)

Innovation is a broad topic. This complex activity includes numerous dimensions:

– Traditional dimensions: strategy, marketing, technology, finance, legal matters, management, etc.

– More specific dimensions: creativity, intellectual property, the psychology of creators, etc.

This is the reason why the terms used for innovation sometimes have quite different meanings depending on the, highly varied, actors that use these terms. Some people even talk about a "real semantic mess", which greatly disturbs cooperation between these actors!

It therefore seems important to define the key terms of innovation that will be used in this chapter. The first definition to be settled is of the term "innovation" or rather, to prioritize action, the verb "to innovate". To innovate is to introduce something new into a field [KER 01, KER 09]. This definition is very close to the definition given in traditional dictionaries, such as, "to introduce something new to an established system" [COL 12].

This definition has several advantages:

– it can be used by all and for all innovations, in both professional and personal contexts;

– it summarizes the "method" that must be used to innovate;

– it is entirely consistent with the definitions of other terms (field, discovery, idea, creation, invention, etc.) used for innovation and intellectual property.

6.1.1.1. *From idea to innovation…*

An idea is a proposal for innovation. This idea leads to a technical creation when it is sufficiently formalized for people "in the field" to be able to make it a reality. If this technical creation is a "world premiere", then it is an invention, which can in turn contribute to the development of one or more innovations.

Technical creations and innovations thrive on ideas, and also, especially, on numerous inventions of all ages (the wheel, the screw, molding, the battery, the electric motor, Teflon, GPS, the Internet, etc.).

At the heart of creation is design. Designing (something) is to achieve the mental creation (of this thing). This definition highlights the very strong links between design and intellectual property.

6.1.1.2. *Using intellectual property… to innovate*

The patent (of an invention) is an intellectual property right that protects technical inventions.

Intellectual property (IP) is everything that relates to the ownership of intellectual creations (IP code, the IP rights of a firm, IP actors, the IP management of a firm, IP economics, etc.).

A firm's IP management is the management of all aspects of IP that are included in that firm's operations and activities (IP surveillance, contracts for innovative collaboration, studies on freedom of usage, patent creativity, acquiring and exploiting IP rights, IP disputes, IP taxation, IP skills, etc.).

6.1.2. *Links between innovation, inventions and patents*

Technical innovation is fertilized by inventions and patents. Figure 6.1 describes the highly valuable relationships between these three domains, and, more generally, between innovation, technical creations and intellectual property.

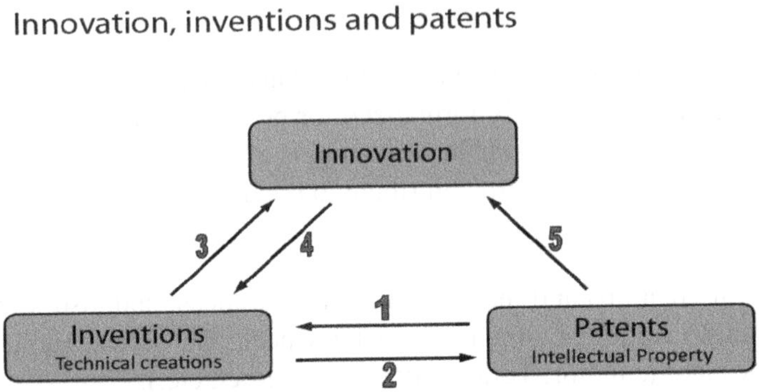

Figure 6.1. *Innovation, inventions and patents (source: author)*

6.1.2.1. *Link: Patents to Inventions*

Existing patents are very useful for technical creations. In fact, all technical solutions created today only combine in a new manner, in a given field, means that can be found for free in the global patent database! This may appear surprising but, for the time being, creators have not yet found an exception to this "rule". The author thanks in advance any reader in a position to give him any information about any exception.

These technical creations are inventions if they are new from a worldwide point of view.

6.1.2.2. *Link: Inventions to Patents*

Inventions are patentable, if they satisfy three patentability criteria defined in the Intellectual Property Code: novelty, inventive step and industrial application.

For designers, the only truly important criterion is "novelty". The "industrial application" criterion is always approved, except in certain very specific cases (relating to business methods, etc.). The "inventive step" criterion, meanwhile, is quite subjective. Only a specialist in patents is really able to understand and file the patent application "to make a nice inventive step appear" (as per the traditional saying among specialists); it is the art of advocacy!

Patents protect inventions. More generally, intellectual property offers protection to most technical creations under various forms (patents, publications, secrecy, contracts, etc.).

6.1.2.3. *Link: Inventions to Innovation*

Inventions feed directly into technical innovations and, indirectly, into other types of innovations.

The invention of the computer mouse illustrates this well. The famous "mouse" has allowed a very large audience to use a computer, which has indirectly contributed to the development of numerous innovations in services: access to databases, electronic mail, e-commerce, etc.

6.1.2.4. *Relationship: Innovation to Inventions*

Innovations, during their development and exploitation stages, generate new needs and new opportunities. This is the primary source for generating inventions.

Take this example from everyday life. In around 1815, a German, Karl von Drais, invented the forerunner of the bicycle, the *draisienne* or dandy horse. The user sits on a seat and his feet rest on the ground in order to move … and brake! The velocipede, and, later, the bicycle have caused the dandy horse to fall into obscurity, but it made a comeback several years ago for children learning to ride their bikes, which in turn gave the inventor the idea to file a patent for a double-function machine, a combined dandy horse and scooter. By turning part of the frame by 180°, the dandy horse becomes a scooter and vice versa (patent no. WO 2013/090953 A1).

6.1.2.5. *Link: Patents to Innovation*

When used correctly, patents facilitate cooperation and the peaceable and profitable exploitation of innovations.

When it is unclear who owns what, cooperation is not possible. The French car manufacturing group PSA, for instance, receives more than 1,000 "unsolicited inventions" every year. These come from people outside PSA who propose innovations in the automobile industry. PSA immediately asks the people proposing these innovations one question: "Have you filed a patent?" "If this is the case, we can look at what you have initially proposed and see if it interests us. If this is not the case, we do not want to know what you have proposed or discuss it. There is a high probability that we will have disputes and that you will accuse us of having stolen your ideas!"

These comments about the links between inventions and patents highlight the fact that the power of innovation lies mainly in the quality of interactions between technical creations and patents.

The language of patents[1] is at the heart of these interactions. This language is logical and powerful, but it does not come naturally to designers and, more broadly, to innovators; this is the reason for interest in the "means tree", which makes the language of patents accessible to all.

6.2. The means tree: a language for innovation and good use of patents

6.2.1. *Is the "language of patents" accessible to all?*

The patent system, created over two centuries ago, is based on a remarkable principle.

A patent is a "contract" between an inventor (mostly representing a company) and the community. The inventor explains how his invention can be achieved. In exchange for this invaluable information, the inventor can be granted a monopoly on exploitation of his/her invention, for a maximum period of 20 years.

In order for this contract to truly be "win-win", the designers must be able to read patents "quickly and efficiently". However, until now, relatively few designers and innovators have this skill.

Are designers and innovators incapable of understanding the "language of patents"?

1 A rather "specific" language found in patents.

6.2.2. *The means tree: the "language of patents" translated into graphic patterns*

The texts of the first patents consisted of traditional technical descriptions. Inventors could therefore easily read the patents.

Over time, the patent offices imposed editorial rules that allowed them to respond more effectively to related key legal questions:

"What are the limitations of the protection of this patent?"

"Does this other technical solution infringe this patent?"

Countries have progressively adopted the use of claims in the texts of patents. Thanks to claims, potential inventors have access to the core of a technical solution, for claims ask them the following crucial question: "What new combinations of means are proposed?"

Patent specialists have therefore invented a remarkable protective language, which is also a powerful design language, because designing consists of combining means to meet a need.

However, designers often have difficulty understanding the claims, which are complicated and often use very long sentences.

On the contrary, if we represent the claims as diagrams, called "means trees", the "language of patents" becomes very clear to designers.

Here is, for instance, a means tree corresponding to the first claim on patent FR 2 752 374, entitled "Device for raising products, particularly condiments, from a receptacle".

In Figure 6.2, the drawings contained in the patent file, display a single raising device (drawing on the left) and a dispensation device in a jar (drawing on the right).

Figure 6.3 shows Claim 1 of patent FR 2 752 374: a sentence made up of 15 lines, and difficult to understand for people who are not specialists!

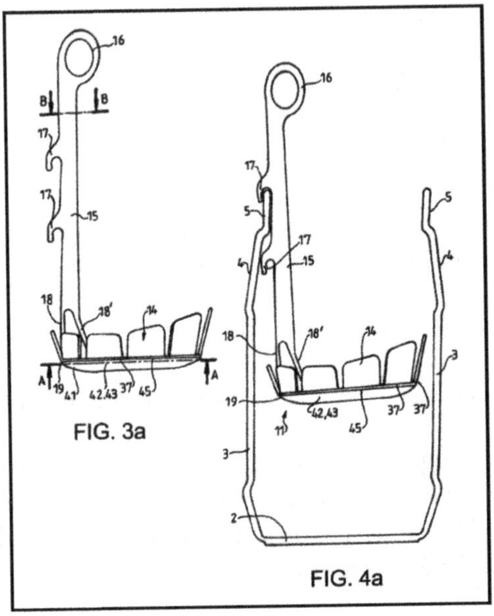

Figure 6.2. *Drawings of patent FR 2 752 374 (source: patent FR 2 752 374)*

FR 2 752 374: Raising device *(Claim 1)*

1. Raising device, including a tray with a contour preventing products from dropping when it is immersed in a receptacle, as well as at least a single independent rod, at one of its ends of the tray and fitted with a gripping end able to rise the tray along with the products in the receptacle, with the rod (15, 21, 22) being at least on part of its length off-centre relative to the tray (11) in such a way as to slip along the edge of the receptacle (1) during its immersion, characterized in that the rod (15, 21, 22) has at least an adjustment device (17, 25, 27) making one piece with the rod and including at least a gripper 27) allowing the rod to grip on the edge of the receptacle according to a given raising position.

Figure 6.3. *Claim 1 of patent FR 2 752 374 (source: patent FR 2 752 374)*

Figure 6.4 shows, in the form of a means tree, the claim presented in Figure 6.3.

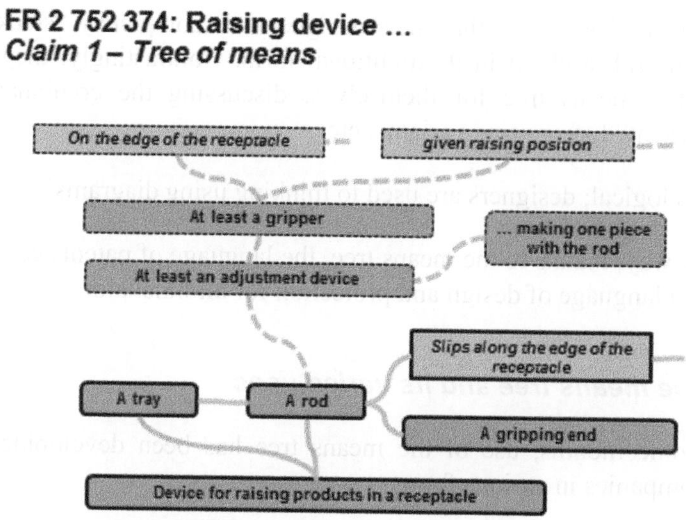

Figure 6.4. *Claim 1 representation using a means tree (source: author). For a color version of this figure, see www.iste.co.uk/uzinidis/innovation.zip*

The following elements are used to draw a means tree:

– a blue rectangle for means;

– an orange link for a time-independent relationship;

– an orange arrow for a time-dependent relationship (process);

– an orange rectangle on the link or the arrow where the relationship is described;

– a "solid line" for means or relationship from the existing "reference" solution;

– a "dotted line" for new means or relationship with respect to the existing "reference" solution.

Thanks to this means tree, it is easy to understand, in the above example, that:

– raising devices already existed, including a tray and a rod slipping along the edge of the receptacle;

– the invention consists of adding hooks on the rod to grip it on to the edge of the receptacle.

Experience has shown that most designers are completely "stuck" when confronted with a claim in its traditional form. Contrastingly, they swiftly claimed the means tree for themselves, discussing the combinations of means presented, proposing variants, etc.

This is logical; designers are used to thinking using diagrams!

In this way, thanks to the means tree, the language of patents can become a common language of design and protection for all innovators.

6.2.3. *The means tree and its varied uses*

For some months, use of the means tree has been developing rapidly within companies in various forms:

– training in patents and intellectual property (IP) use;

– uses of existing patents (patent surveillance, patent extensions, etc.);

– brainstorming sessions based on patents;

– preparation for filing patent applications, etc.

6.2.3.1. *Training in the use of patents and intellectual property (IP)*

Most innovative technical companies have set up training programs so that various actors, particularly, designers, know how to use patents and are familiar with IP rules.

Training can be boosted thanks to the means tree, as it enhances sessions dedicated to:

– reading patents;

– creativity based on patents;

– filing patent applications;

– cooperation between designers and IP specialists.

6.2.3.2. *Use of existing patents*

The means tree allows various actors (decision makers, designers, etc.) to better understand what is protected by each patent. This in-depth understanding is very useful:

– For making good use of surveillance and patent state-of-the-art analysis:

- patent analysis;

- sessions for exploiting the selected patents.

– For fostering technical expertise (led by technical referrers):

- state-of-the-art appropriation;

- appropriation of solutions protected by the company and by third parties;

- design the appropriation of key factors traceability.

– For shedding light on the decisions taken, by properly integrating the three aspects (strategy, technology, legal aspects) for:

- framing innovation projects;

- project partnerships;

- studies on freedom of usage;

- filing, extending and abandoning patents;

- analyzing potential infringement;

- exploiting patents, etc.

6.2.3.3. *Creativity brainstorming sessions based on patents*

The means tree points out the means combination at the heart of technical solutions, whether they are patented or not.

From these means combinations, it is easy to stimulate creativity through:

Substitution: by replacing one means with another means, a relationship between means with another relationship, etc.;

Transfer: by identifying means and relationships, which are used in other technical fields, and by trying to transfer them to one's own technical field.

Moreover, the means tree can be used to formalize and "challenge" the ideas for solutions generated by all creativity methods (brainstorming, TRIZ, etc.).

6.2.3.4. *Preparation for filing patent applications*

The means tree offers designers and patent specialists (patent engineers, intellectual property consultants) a better mutual understanding. It therefore provides companies with important advantages:

– improvements in the quality of the patents filed;

– saving time for designers and patent specialists (therefore reducing costs);

– reductions in the preparatory phases of filing patents;

– improvements in the quality of the relationship between designers and patent specialists which is universally appreciated.

6.2.4. *The means tree: a "turbocharger" for the "innovate thanks to patents" approach*

All uses of the means tree are part of the comprehensive approach called "innovate thanks to patents". This approach departs from the idea that patents are not only a "legal outcome" of innovation projects, but also useful at all stages of the innovation process.

In fact, patents are simultaneously:

– a strategic tool for protecting and exploiting innovations;

– a mine of information;

– a powerful design language.

Thanks to these three complementary functions, patents can illuminate, foster, secure and better utilize innovation projects.

The two following diagrams show:

– the main advantages linked to use of patents;

– the patent activities that allow these advantages to be created.

In these two diagrams, the advantages and activities are positioned in relation to stages in the innovation process. The size of the headings reflects their importance to a company that has been taken as a model. In every company, the importance of advantages and activities depends, naturally, on the decisions made by its managers.

Figure 6.5 distinguishes two kinds of advantages: advantages linked to filing patents (boxed headings) and advantages linked to "free" use of patents.

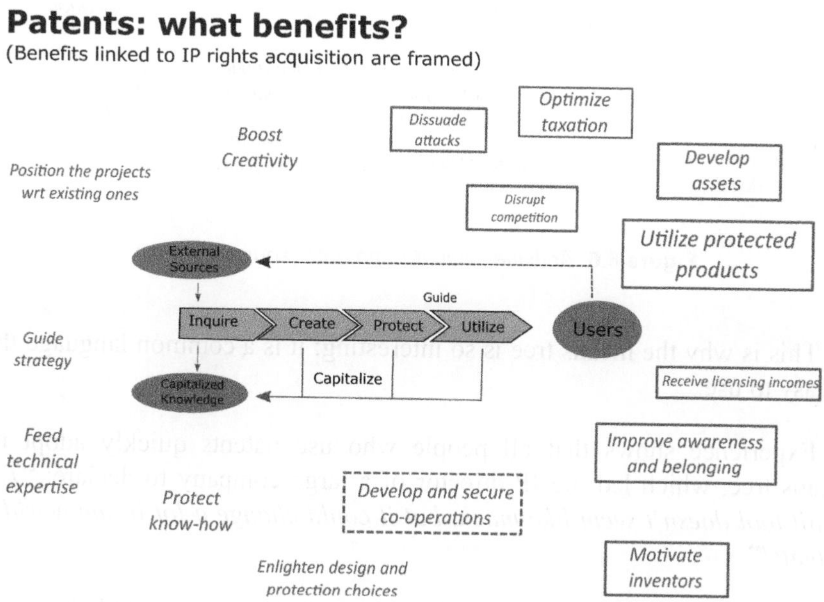

Figure 6.5. *The advantages of patents (source: author)*

Figure 6.6 shows that "patent activities" are numerous and often interconnected. The usage of patents is therefore both rich and complex. The advantages gained by companies thus directly depend on the quality of cooperation between patent specialists (within the company and/or outside it) and non-specialists.

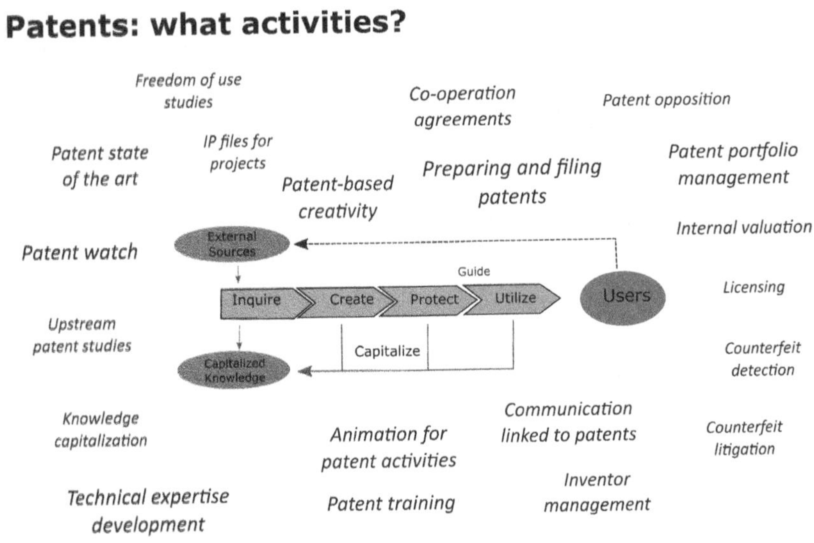

Figure 6.6. *Activities linked to patents (source: author)*

This is why the means tree is so interesting: it is a common language that is easy to use.

Experience shows that all people who use patents quickly adopt the means tree, which led the IP director of a large company to declare: "*This small tool doesn't seem like much, but it could change a lot in the world of patents!*"

Furthermore, beyond the new direct uses of patents presented above, the patent system can serve as a model to foster knowledge management dedicated to innovation.

6.3. The patent system: a model for the management of knowledge for innovation

As has already been shown (section 6.2.1), the patent system has been created to accelerate technical progress, thanks to a relevant acknowledgment of inventors and effective knowledge sharing.

Its aim is very close to that which all companies increasingly seek: to motivate innovators and share knowledge better in order to be able to innovate more effectively [ERM 03, SAU 13].

For more than 200 years, the patent system has proved its effectiveness in this field. Millions of innovators working in varied companies around the world can freely and easily access over 80 million documents. These documents are very rich; they document the history of inventions over more than two centuries, across all technical fields.

The fathers of the patent system, particularly Benjamin Franklin and Thomas Jefferson (two inventors who participated in writing the US Constitution after the War of Independence), could not have imagined, in 1790, the revolution that information technology would bring at the end of the 20th Century. Moreover, when the patent system was computerized, nothing changed in the structuring of knowledge in this system (structuring patents, structuring patent databases, processes). The goldmine that is the global patent database has existed for decades. Today, thanks to information technology, all innovators can exploit it.

In addition, since 1790, it has never been necessary to clean up the 80 million documents in the global patent database. What company can say such a thing about its intranet devoted to innovation (much more recent and with much fewer documents!)? What is the secret behind this performance?

The secret is simple and all companies can draw inspiration from it. The structuring of patents focuses these documents on the essential element: the traceability of design choices. In this way, each patent, from the moment it is filed, generates a new state of the art and proposes a new technical solution that improves the existing one. A patent therefore does not obstruct the patents that follow it. On the contrary, it is a link to the history of technology.

This patent structuring is thus highly effective in capitalizing on knowledge with the aim of fostering innovation as the description of a patent corresponds to the list of questions that must be asked to innovate.

The patent's claims cut straight to the heart of the design by asking the crucial question: "*What are the new combinations of the proposed means?*"

Patent outline: ***the underlying questions***

- **Description**
  - **Technical field of the invention**
    - *In which field is the invention?*
  - **Background of the invention introducing the issue problem**
    - *What is the problem to be solved? How has it been solved up to now?*
  - **Summary of the invention**
    - *What are the new proposed solutions: purpose, technical means, advantages ...?*
  - **Brief description of the drawings**
  - **Description of at least one embodiment**
    - *How will the new solutions be implemented?*
  - **Utility of the invention**
    - *What are the potential applications of the invention and their utility?*
- **Claims**
    - *What are the new combinations of means proposed?*

Figure 6.7. *The underlying questions about the patent (source: author)*

This remarkable patent structure, shown in Figure 6.7, has given rise to two very useful forms for innovation management and innovators: the "innovation memo" and the "knowledge sheet". These two forms are simple, powerful and universal tools. Companies use them for all types of innovations (technical products, proposed services, internal organization, etc.) and all people can also use them in their personal lives (to improve tidying the basement, to get on better with next-door neighbors, deepen knowledge of foreign languages, etc.).

It really works for all areas!

The innovation memo presented in Figure 6.8, inspired by patents, takes on and adapts the questions that must be asked in order to file a patent application. The questions on the second page extend the logic behind patents and allow knowledge to be shared further within a company, particularly for innovation projects, with questions such as:

– What are the restrictions/disadvantages of the chosen solutions?

– What other solutions have been conceived/studied? Why have they not been chosen?

Innovation Memo

Author Name	Unit	Tel	E-Mail
Date:	N°:	Category:	Status:

Title

1. Innovation field

What is this about? In which field(s) can the innovation be categorized?

2. Need and existing solutions

What need is to be met? What is to be improved? What is the issue?
How is this need currently satisfied?
What are the existing solutions in the field of innovation? What are their disadvantages? What are their limitations?
What solutions exist in other fields for similar needs?

3. Proposed solutions: principles

What new solutions are envisaged or proposed (aims, main features, proposed new means combinations)?
What are the advantages and disadvantages of these new solutions, in comparison to known solutions?

4. Proposed solutions: implementation

How are or will be these new solutions (used means, operation, results, etc.) implemented for the basic solution and its variants?

5. Users

Who uses or could use the innovation? To do what? With what results?

Additional information

1. Context in which the innovation was generated

In which context was the innovation generated (innovation project, cooperation with another unit/company, meeting, own home, etc.)? Who knows about it? In what form (transmitted documents, confidentiality level, etc.)?

2. Sources of information used

What are the main sources of information that have been used to achieve the "state of the art" analysis? What are the methods used (key words, etc.)? What results have been obtained?

3. Innovation progress status

What is the progress status of the innovation (descriptive note, expertise, attempt, ongoing project, operational implementation, etc.)? What are the related documents?

4. Limitations of chosen solutions

What are the limitations/disadvantages of the chosen solutions? What are the problems to be solved?

5. Rejected solutions

What other solutions have been conceived/studied? Why have they been rejected?

6. Proposed developments (at the end of the innovation project)

Must developments be envisaged? Which ones?

Figure 6.8. *The innovation memo (source: author)*

The knowledge sheet shown in Figure 6.9, inspired by the patent summary (the first page of the patent), is a kind of "billboard ad" for the corresponding knowledge module. The knowledge sheet has, like the patent summary, two main functions:

to present an overview of the knowledge module (innovation memos, expertise, training support, instructions for machine maintenance, website presentations, etc.);

to facilitate the research of relevant knowledge modules and the establishment of contact with appropriate people in the field.

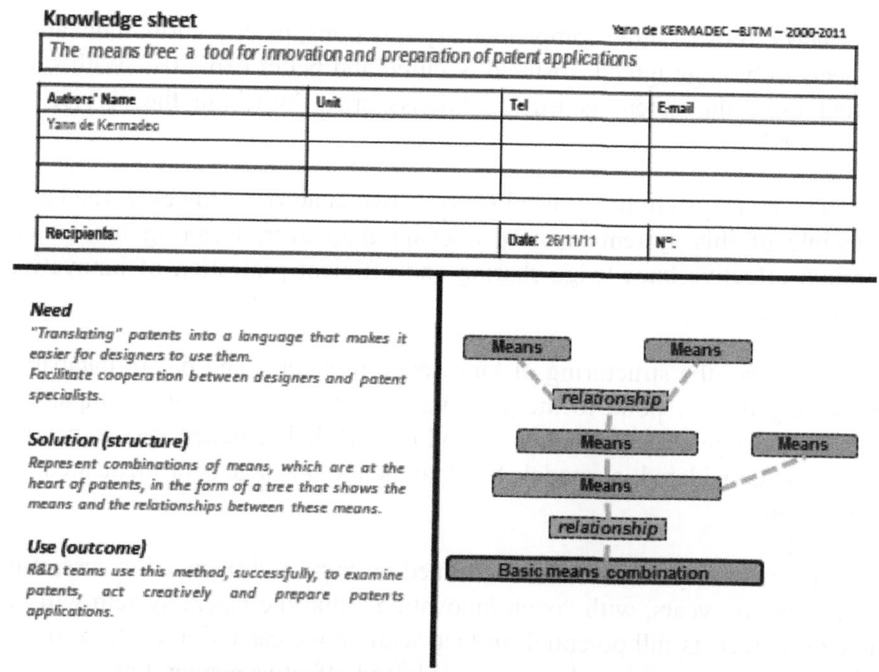
Knowledge sheet

Yann de KERMADEC –BJTM – 2000-2011

The means tree: a tool for innovation and preparation of patent applications

Authors' Name	Unit	Tel	E-mail
Yann de Kermadec			

Recipients:	Date: 26/11/11	N°:

Need

"Translating" patents into a language that makes it easier for designers to use them.
Facilitate cooperation between designers and patent specialists.

Solution (structure)

Represent combinations of means, which are at the heart of patents, in the form of a tree that shows the means and the relationships between these means.

Use (outcome)

R&D teams use this method, successfully, to examine patents, act creatively and prepare patents applications.

Figure 6.9. *The knowledge sheet (source: author)*

This example of a knowledge sheet, which shows the "means tree", allows us to conclude by returning to an innovation at the heart of this chapter.

Indeed, the means tree is the language that allows proper use of the patent, the "Swiss army knife" of innovators, which is highly efficient for cooperation, invention and innovation.

6.4. Conclusion

This chapter highlights the fact that patents are simultaneously:

– a strategic tool for protecting and utilizing innovations;

– a mine of information;

– a powerful design language.

Thanks to these three complementary functions, patents allow innovation projects to be highlighted, fostered, secured and better exploited. With such varied uses, the patent is truly a "Swiss army knife" in the service of innovators!

The patent system has existed for over two centuries. However, the basic principle of this system is more relevant than ever, fostering innovation through effective knowledge sharing and efficient protection of innovative companies.

Over time, the structuring of knowledge in patents and patent databases has changed to respond to the growing demands of states and companies. This knowledge structure is logical and powerful. The patent system can be presently considered a model for knowledge management dedicated to innovation.

If the patent system is not undermined by legal and financial excess, it is in the coming years, with "open innovation", that the patent system will be able to deliver its full potential. In fact, without a clear vision of "who owns what", it is not possible to have peaceable and effective cooperation.

Moreover, the wealth of technical information, freely accessible in databases, is still under-utilized. As soon as this goldmine is properly used, the patent usage barycenter will, naturally and healthily, be at the heart of design and innovation.

However, in order to do so, designers must be comfortable with patents, and therefore understand the language of patents and the power of this language in design. The means tree, which has been presented in this chapter, could greatly contribute to this, particularly if the young people undertaking further technical education claim it and use it during their studies.

Indeed, if students use the means tree during their end-of-course internship, they will quickly become patent lovers, and transfer this passion, and the related skills, to people with whom they work in companies.

6.5. Bibliography

[ALT 00] ALTER N., *L'innovation ordinaire*, Eyrolles, 2000.

[BEF 15] BEFFA K., VILLANI C., *Les coulisses de la création*, Flammarion, 2015.

[BRE 02] BREESÉ P., *Straégies de propriété industrielle*, Dunod, 2002.

[CCI 13] CCI-PARIS, BETA-Strasbourg, L'innovation ouverte – Définition, pratiques et perspectives, Chambre de commerce et d'industrie de Paris, 2013.

[COL 12] COLLECTIF-ROBERT, *Le Dictionnaire historique de la langue Française*, Le Robert, 2012.

[DRU 15] DRUCKER P., HESSELBEIN F., *Les cinq questions fondamentales du management*, Diateino, 2015.

[ERM 03] ERMINE J.L., *La gestion des connaissances*, 2nd ed., Hermès-Lavoisier, 2003.

[FON 10] FONTANET X., *Si on faisait confiance aux entrepreneurs*, Les Belles Lettres, 2010.

[GEN 01] GENELOT D., *Manager dans la complexité*, 3rd ed., INSEP Editions, 2001.

[GOM 13] GOMEZ P.Y., *Le travail invisible*, François Bourin Editeur, 2013.

[KER 01] KERMADEC Y. DE, *Innover grâce aux brevets*, 2nd ed., INSEP Editions, 2001.

[KER 09] KERMADEC Y. DE, *Innover, c'est l'affaire de tous*, 2nd ed., INSEP Editions, 2009.

[KER 10] KERMADEC Y. DE, *PME: pensez propriété intellectuelle!*, DGCIS, 2010.

[KER 15] KERMADEC Y. DE, BREESÉ P., *La propriété intellectuelle au service de l'innovation*, 4th ed., Nathan, 2015.

[NOY 02] NOYE D., *Réussir les changements difficiles*, INSEP Editions, 2002.

[PMI 00] PMI, *Management de projet – un référentiel de connaissances*, AFNOR, 2000.

[RAD 15] RADJOU N., *L'innovation frugale*, Diateino, 2015.

[ROU 03] ROUXEL J., *Les Shadocks*, Editions Circonflexe, 2003.

[SAU 13] SAULAIS P., Application de la gestion des connaissances à la créativité des experts et à la planification de la R&T en milieu industriel de haute technologie, Doctoral thesis, Télécom Ecole de Management, 2013.

7

Invention, Innovation and Intellectual Property Rights

> "Science and technology have truly sealed their union over the course of the closing century [...] but it was only twenty years ago that large companies, driven towards creativity by competition, came closer to the universities as soon as an idea for a new production or a new method of production was born. Universities, coping with reductions in public R&D spending, saw in value-creation, in research, through contracts with companies or by filing and exploiting patents, a means of financing their activities [...] On the other hand, the public funding of research has long been justified by the fact that the knowledge it produces has consequently been widely and freely disseminated" [LAP 01a].

In this way, this author notes the increasing transformation of intellectual creations into marketable goods and shows his interest in the roles played by intellectual property in the different life stages of an innovative company. The first links established between technical creation, industrial property rights and innovative businesses are thus largely economic, with only the industrial property taking action. The success of an invention is not only dependent on the creativity of its inventor or on the unique filing of a patent application, but also (and especially, from this economic perspective) on a range of economic, social and political factors that encourage or penalize the diffusion of inventions [BOU 01]. In addition, Dimitri Uzunidis [UZU 01]

Chapter written by Pierre SAULAIS.

focuses his attention on "the marketability of public research", as seen in the French law on innovation and research of July 12, 1999 (Loi no. 99-587), in terms of the private appropriation of public research results, while the objective of this law is "to intensify the systemic relationships between public research, entrepreneurship and innovation" [UZU 01].

Having established this, it seems to us that, apart from the economic prism and its analysis of effects, few academic works have tried to bring together the knowledge areas represented by the study of the founding principles of inventing, the study of the founding principles of innovation and the study of the founding principles of legal property [SAU 13]. Our intention here is therefore to consider these three elements – innovation, invention and legal property – not only as three economic data, but also as three objects of knowledge with the aim of exploring their nature, and then their genetic interactions (in the sense of the generation dynamic).

This chapter outline follows directly from our above-mentioned intention. In the first part (sections 7.1, 7.2 and 7.3), we will endeavor to present each of the three knowledge objects succinctly and successively, beginning with the act of innovation and its results, followed by the capacity to invent and finally the intellectual inventive corpus and intangible capital (which we will take great care to distinguish). The second part (section 7.4) will be dedicated to providing an overview of the links between these three knowledge objects, whose combinatory is worth six. This will only be an overview as some elements of this combinatory are described in greater depth and in a specialized way in other chapters. In the third part (section 7.5), we will question the nature of these links insofar as one of the knowledge objects is able to stimulate another, with our questioning concentrating on a possible cause of this capacity.

7.1. Innovation

7.1.1. *The multidimensional aspect of innovation*

According to Peter Drucker [PRA 05], innovation consists of the determined and organized quest for change and the systematic analysis of the opportunities that such changes can offer in terms of economic and social progress; innovation is the act of assigning resources a new capacity to create wealth and to disseminate it. An innovation could therefore be an invention that has been accepted by the market, so that a new idea which has been produced but not exploited, and which has not materialized in the form of a product or service will only remain an invention. Jean-Yves Prax [PRA 05] emphasizes that innovations do not appear by accident, but are the product of the market's capability to appropriate new uses, by creating a dialectic between technical progress and social progress. The impact of the change brought by innovation to economic activity is an important part of recent works of research in the economics of innovation [GUE 09].

A wide range of perspectives allow us to shed light on the term "innovation". The concept of innovation at once describes an outcome, an offer of products and/or services, or a change in a production process [SOP 07]. It refers to multiple fields of research falling within the areas of corporate strategy, marketing and project management [SOP 07]. It is therefore a polysemy, which varies according to context and can indicate either the creation process leading to a new element with an economic and social value, or the element itself [FER 06]. These three meanings suggest that innovation only exists in relation to the entity that accepts it, that it involves dissemination and that it is a deliberate act, which leads to the following definition:

"We define innovation as a deliberate organizational process that leads to the proposal and adoption, on the market or within a company, of a new product (by the guidelines of AFNOR, the French national organization for standardization). This process allows one or several companies to improve their strategic position (to gain or increase their market power) and/or strengthen their skills or their technologies. The new product can be a physical object, a service, a technology, a new skill or a combination of several of these variables" [FER 06].

Innovation is thus not only technological, organization or commercial; it in fact integrates all of these different dimensions.

7.1.2. *Innovation procedures and processes*

7.1.2.1. *Innovation procedures and processes according to the industrial state-of-the-art*

A recent study by the *Club Gestion des Connaissances*[1] synthesized numerous innovation operating methods in companies and shows that all of

1 The *Club Gestion des Connaissances* (translation = knowledge management club) is an association of companies founded in 1999 with the aim of building a set of common references for concepts and pragmatic tools in order to implement knowledge management in companies.

these methods follow more or less the same procedure, encapsulated in Figure 7.1 and detailed in Le Loarne and Bianco [LE 09]:

– the strategic pre-positioning assumes innovation to be a mode of competition;

– the generation of ideas relies on stimulating creativity. The construction and qualification of concepts consists of transforming the ideas into innovation concepts;

– the selection of innovation projects is made according to the choice of presentation mode for the projects following the definition of internal selection criteria;

– the deployment relies on the protection, dissemination and value creation of the innovation.

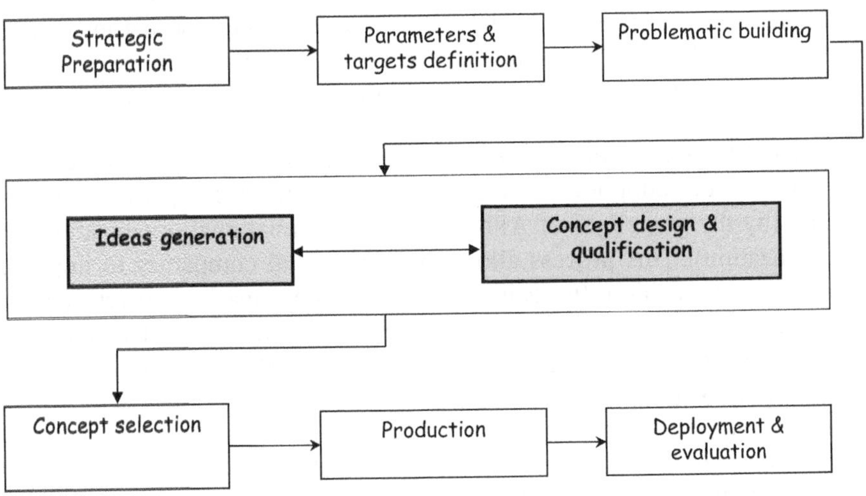

Figure 7.1. *The innovation process (source: Pierre Saulais)*

What all of these methods have in common is the generation of ideas, which the left part of Figure 7.2 shows under the heading "Front End of Innovation", for an innovative market-oriented approach.

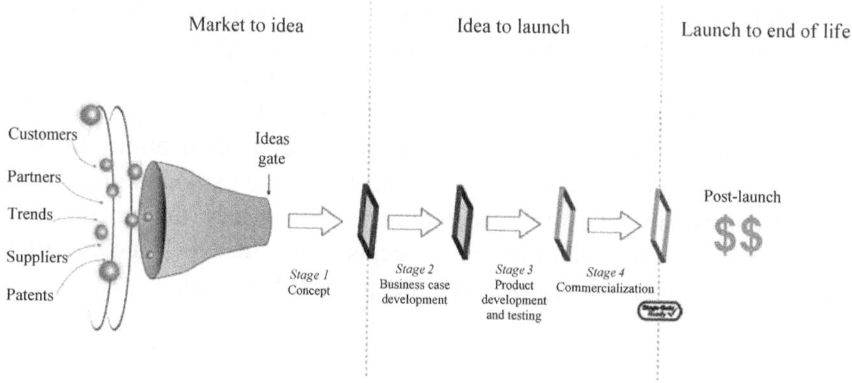

Figure 7.2. *The Front End of Innovation (source:* Club Gestion des Connaissances*)*

At this stage, we must recall:

– the concept of change, implying progress and having to be the object of market appropriation;

– the result, product or service or process or new mode of organization, effectively sold or implemented;

– the focal point of idea generation (or ideation).

7.1.2.2. *Innovation procedures and processes according to the academic state-of-the-art*

7.1.2.2.1. Variety of perspectives on the innovation process

The innovation procedure is marked by "a dual nature of recurrence and uncertainty and its management requires vision and creativity" [FER 06]. According to these authors, there are eight points of view which can be mentioned in the rolling out of the innovation process: value creation in technical progress [BIA 74], the adoption of a novelty [ALT 00], a whirling process [CAL 85], a marketing sequence [LAM 86], the transformation of a technical system [MAF 98], a project [DEC 80, NAV 89] and a collective learning process [HAT 94]. The innovation process can therefore represent "a collective creation, organized in time and space, with a view to demand" [GIA 93], "an open heuristic which pits against each other, on the one hand, individuals open to the ultimate aims, projecting values and representations, and, on the other hand, a physical and social context transformed by the intervention, but which answers, surprises and transforms the trajectory of the designer in return" [GAR 95], a process of "creating meanings and new knowledge, which are most often tacit" [CHA 99], a deliberate organizational procedure representative of innovation situations that require the articulation of innovation activities and exploitation activities, ensuring consistency over time, according to four phases: the emergence of innovative ideas, the maturation of concepts, the effective launching of an innovative project and the realization of the project [SAU 13].

7.1.2.2.2. The multiple dimensions of managing the innovation process

Few works address the different aspects of innovation [TID 06] in an integrated way: according to those authors, companies are inevitably limited in their choice of innovation strategies by their accumulated abilities as well as the opportunities they are able to exploit, meaning that they are situated on technological trajectories.

7.1.2.2.3. A global innovation process involving an approach based on organization and knowledge management

Debra Amidon [AMI 01] finds the power of innovation in intellectual capital and considers this power to result from the combination of rapid knowledge sharing and the evolution of new applications. The author places each of these dimensions into the knowledge economy, where the manageable assets are formed by knowledge and by all associated intangible values, and where the architecture of the innovation management system is designed with the aim of optimizing the use of financial, human and technical resources: performances in the knowledge economy, structures for the exploitation of knowledge, individuals as knowledge actors, knowledge-processing technology and the knowledge management process. Their point of view is taken up by Benoit-Cervantes [BEN 08], according to whom the innovation process can be illustrated by four aspects (Product, People, Processes and Cultural Potential) [BEN 08] (from Higgins).

The academic references emphasize two key points for innovation: the necessity of change and learning. We can thank Gregory Bateson for having provided the key elements for understanding the process of change [BAT 72].

For Soparnot and Stevens, organizational learning is the common characteristic of companies with long-term performances, as they have developed a capacity to improve and continually renew their operating methods and their interpretation schemes [SOP 07]. It is interesting to note that innovative companies are also learning companies; innovation projects and organizational learning influence each other reciprocally. Two complementary approaches account for the learning modes. For March [MAR 91], learning is based on the exploitation and exploration processes; exploitation can be summarized by the idea of "learning to do the same thing better" (there is a connection to incremental innovation, which stimulates exploitation). Exploration can be summarized by the idea of "learning to do things differently" (linking to breakthrough innovation, which stimulates exploration). Argyris [ARG 93] and Senge [SEN 91] developed a perspective according to which a company is capable of providing itself with an organizational set-up and with specific mechanisms allowing it to modify its knowledge corpus. Their approach is based on a very specific scientific position: research-intervention [ARG 93, SEN 91]. The organizational learning processes are directly linked to knowledge corpus and its management within the company.

7.1.3. *The intense knowledge stages of the innovation process*

Figure 7.3 is borrowed from the *Club Gestion des Connaissances*. It shows the distribution of the eight stages of the innovation process by identifying the stages where knowledge is predominant.

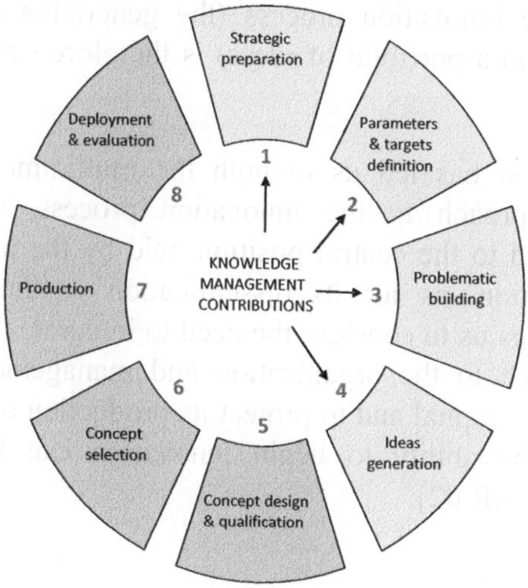

Figure 7.3. *Knowledge contributions in the innovation process (source:* Club Gestion des Connaissances*)*

The most intense stages in terms of knowledge are the first four: strategic pre-positioning, defining and deciding on targets and perimeters, constructing issues and generating ideas. This finding has proven to be fully in line with the introduction, noted by Debra Amidon, of an approach to knowledge management organization in the innovation process.

In addition, as recalled by Soparnot and Stevens, innovation itself contributes to the creation of different kinds of knowledge, which can change over time by combining and interacting with each other [SOP 07]. This is what Nonaka and Takeuchi call the knowledge spiral [NON 97].

Soparnot and Stevens explain that if innovation contributes to the renewal of an organization by boosting knowledge creation, then it is also equally dependent on it, as it develops by means of the organization's accumulated knowledge; innovation management therefore invites knowledge management as, in part, knowledge determines innovation just as it comes from it [SOP 07].

Tidd, Bessant and Pavitt confirm that innovation depends on knowledge by combining different sets of knowledge [TID 06]. To this end, they call on the works of Henderson and Clark [HEN 90], which have examined in detail the types of knowledge involved in different innovation categories.

The core of the innovation process (the generation of ideas that use creativity to build up a portfolio of ideas) is therefore strongly knowledge-based.

This investigation has led us to both the multidimensional aspect of the conceptual approach to the innovation process, which requires a systemic vision, and to the central position held by the innovative idea (as much for its creation, as for its dissemination or the difficulty of its assessment). It brings us to consider the need to integrate into the innovation process an approach to the organization and management of knowledge based on intellectual capital and to protect its production through intellectual property rights. The ability to begin innovation can be referred to as "*innovativeness*" [GAR 02].

7.2. Invention and the ability to invent

7.2.1. *The concept of inventiveness*

The literature provides only a few definitions of inventiveness [ROU 73], due to potential confusion with creativity. By taking up the view point of the Palo Alto school [WAT 75], we can say that inventiveness is the ability to change the elements of a system.

7.2.2. *Combining creativity and inventiveness*

Creativity represents the ability to provoke a phenomenal state of mind and inventiveness is the ability to put the creative mechanism resulting from the creative state of mind into action so as to produce a creation or work. Creativity and inventiveness therefore combine in a dual activity prior to the innovation process [SAU 13].

The innovation process in generating ideas is usually described in four stages [LOU 08]: describing the question, producing different ideas, applying the collected ideas to the question asked, and sorting and selecting them.

The usual techniques for producing different ideas aim to produce the greatest possible number of ideas in a limited time by stimulating imagination (Tables 7.1 and 7.2).

Divergent thinking
Suspending judgment
Looking for quantity
Making note of all ideas
Looking for original, surprising ideas
Combining and connecting ideas

Table 7.1. *Divergent thinking (source: Pierre Saulais)*

Convergent thinking
Assess constructively
Improve ideas
Select according to criteria
Criterion: newness
Criterion: tailored solution

Table 7.2. *Convergent thinking (source: Pierre Saulais)*

7.2.3. *Overview of some standard methods*

There is a plethora of traditional tools in this field – including trial and error methods, synectics, questionnaires and morphological boxes [LOU 08] – as well as numerous techniques: TRIZ method [LOU 08], Thinking Process Approach [LOU 08], Creative Thinking [SIM 69], Lateral Thinking [LOU 08], Brainstorming, Mind Mapping [LOU 08], C-K theory [HAT 02], etc.

The comprehensive problem-solving process shown in Figure 7.4 aims primarily to explain the logic of transitions that bridge the gaps left by contradictions [LOU 08]. It begins by formulating a problem statement that is built progressively through the examination of various types of contradictions.

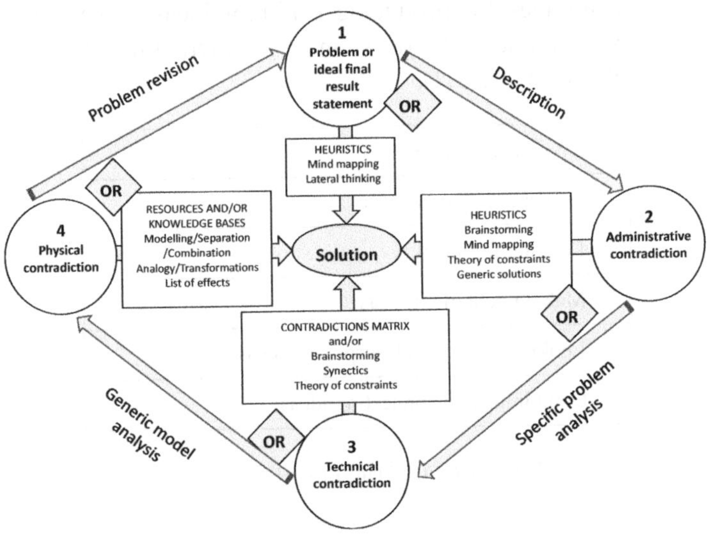

Figure 7.4. *Comprehensive process of integrating creativity techniques (source: [LOU 08])*

The term TRIZ is the Russian acronym for a phrase meaning "theory of inventive problem-solving", a systematic method invented by Genrich Altshuller, based on knowledge and drawing its resources from the compilation of a large number of invention patents [LOU 08]. TRIZ brings together a set of methods and tools related to usage rules in order to discover solutions to technical problems [ALT 84, ALT 98]. This theory, which was consolidated in 1995, is based on the idea that discoveries and inventions are founded on universal principles. The TRIZ-based approach allows new solutions to be developed using existing inventions and knowledge which lies outside the field of study. TRIZ can be applied to research and the development of new generations of products by proposing a set of evolution laws. TRIZ addresses the resolution of difficult problems with a methodology and tools based on resolving conflicts inherent to all technical systems [ELT 06].

The systematic study of techniques and their functions is the foundation of TRIZ. One technique can perform several functions, including the primary function that reflects the reason for the technique's existence. Other auxiliary functions accompany the primary function, such as useful functions, harmful functions (producing undesirable effects) and neutral functions. Contradictions form the central concept of TRIZ (see Figure 7.4) and fall into three categories: administrative contradictions (encountered when the research directions for solutions are unknown *a priori*), technical contradictions (one or several research directions emerge that call on similar ways of thinking through the contradiction matrix and Altshuller's 40 principles) and physical contradictions [LOU 08].

7.3. The inventive intellectual corpus

In this section, we will learn about the nature of the inventive intellectual corpus through the act of inventive creation that produces an intellectual work, without prejudging the technical or artistic nature of the creation. In this way, we will first examine the links between intellectual works and the inventive intellectual corpus, followed by the connections between intellectual work and intangible capital. We will then address the description of the traces left in the inventive intellectual corpus (traces that can be seen in the material support of the intellectual work in question), before we make the connection between intellectual property and the innovation process. We will then take the time to examine the issue of the value of intangible capital.

7.3.1. *Intellectual works and intellectual corpus*

7.3.1.1. *The nature of the inventive intellectual corpus*

We will begin by questioning the concept of "patrimony", derived from the Latin *patrimonium* (family property), which itself comes from *pater* (father of the family, head of the household) [GAF 34], which represents an inherited asset that goes down, in accordance with the law, from mothers and fathers to their children and which is, in general, family property [LIT 86]. Emile Littré also notes that, in a figurative sense, "patrimony" alludes to "that which is considered to be 'patrimonial' property" [LIT 86], which, in a literal sense, means capital. According to Gérard Cornu, patrimony represents, by extension, all of the assets commonly recognized as having a significant value, such as the global cultural patrimony heritage, or patrimony [COR 87]. In the sense of cultural heritage, patrimony follows the idea of heritage according to which certain objects from the past must be preserved to be passed on to future generations. Patrimony is always connected to a legal subject. "Patrimony or Heritage is linked to the aim that has been bestowed upon it, dedicated to conservation and transmission when it is inherited from the past and bequeathed to the future or to the needs of its owner when it is the wealth of the present" [ALL 03]. Our aim is to understand the nature of the inventive intellectual corpus when it is a legacy of the past bequeathed to the future, and then to show that this patrimony accommodates a dormant deposit of value, whose current wealth could be extracted by a structured analysis using knowledge engineering methods.

7.3.1.2. *General information on intellectual property rights and copyright*

Copyright protection is awarded from the moment that there is an original creation, regardless of its genre (literary, musical, artistic, etc.), its purpose (informative, advertising, artistic, etc.), its form of expression (oral, written, etc.) or the merit of its author. In its current meaning under the French law, the term "*droit d'auteur*" corresponds to the term "literary and artistic property" in the French Intellectual Property Code.

The Intellectual Property Code stipulates that the author of an intellectual work enjoys over that work, due solely to having created it, an intangible exclusive property right that includes intellectual and moral attributes, as well as patrimonial attributes (Article L 111-1). This intangible property is, in essence, distinct from the tangible property of the physical support of the intellectual work. Intangible property refers to the content of the work, and is by definition dematerialized and inalienable due to its purely intellectual nature. The tangible property refers to the support of the work, which belongs to the sensitive world, can be disseminated and be the subject of all kinds of transactions, and participates in the information system, considered here to be the repository of media materializing to the creation of an intellectual work.

The intellectual property rights on a work are acquired without any depositing formalities due solely to its creation; it is therefore a *sui generis* law, which is characteristic of its type [BLO 32]. It confers property rights (rights of reproduction and representation) and moral rights (notably naming rights and the right to the inviolability of the work).

Being the right of the author, the "*droit d'auteur*" is the right of the creator; although the designation of "literary and artistic property" is given to the "*droit d'auteur*", it does not bestow property upon the creators in the sense of the civil code, but only grants them a temporary exclusive usage privilege.

7.3.1.3. *Representation of the inventive intellectual corpus*

The starting point for our analysis of the nature of the inventive intellectual corpus is made up of the act of creative invention, which corresponds to the realization phase of the creation phenomenon with two aspects, creativity and inventiveness. The product of this activity (the intellectual work) only generates intellectual property rights if the work satisfies conditions of originality that embody inventive creativity; that the work will enrich a corpus that we call the inventive intellectual corpus. Just as an *a posteriori* characterization of the nature of the work can be achieved by satisfying the criteria for awarding intellectual property rights, these rights are therefore a precious resource of valuable material to learn about the nature of the inventive intellectual corpus. The intellectual creation that is characteristic of the intellectual work is transformed into new knowledge, and it is this new knowledge which, after formalization, goes on to expand the intellectual corpus that forms the inventive intellectual corpus, from a dynamic point of view.

Any intellectual work may produce intellectual property rights which are traditionally separated into industrial property rights (including industrial patents) and "*droits d'auteur*" (including literary and artistic works).

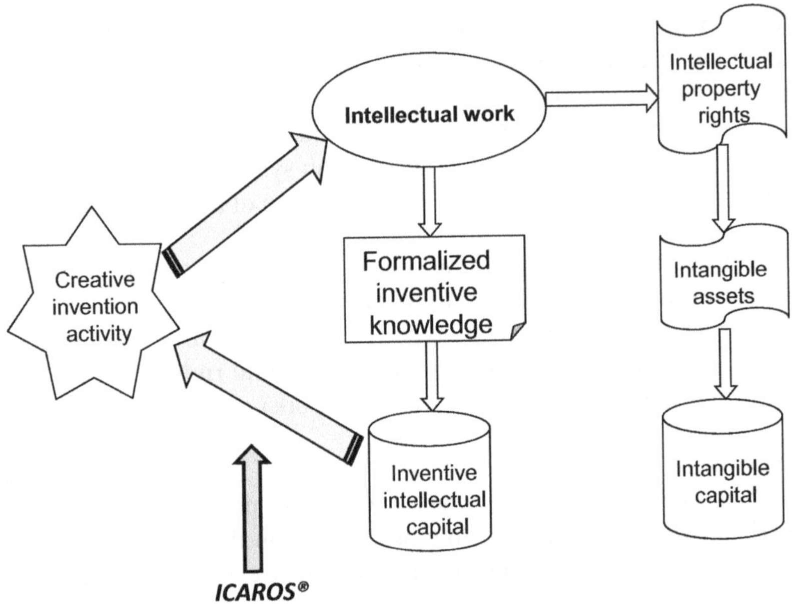

Figure 7.5. *The inventive intellectual corpus (source: Pierre Saulais)*

As shown in Figure 7.5, the act of creative invention is the *deus ex machina* that produces the intellectual work [SAU 13]. Recognizing this work as original (in the artistic and literary field) or as new and inventive (in the technological field) allows, in the technological field on which we are focusing, the intellectual creation characteristic of the intellectual work to be validated. This intellectual creation is expressed in the form of new knowledge which, after formalization, goes on to dynamically enlarge the intellectual corpus that forms the inventive intellectual corpus. We have seen how inventive intellectual creation, by taking its place in the inventive intellectual corpus (through the new knowledge that it expresses), defines its nature and suggests its method of analysis. The inventive intellectual corpus is therefore none other than the dematerialized deposit of this new knowledge, fertilizing the previous knowledge accumulated in the existing corpus of knowledge, which is at once the legacy of the past and the wealth of the present.

The characterization of the nature of the work can be obtained *a posteriori* by satisfying the criteria for awarding intellectual property rights that consecrate the creator as the legally and economically recognized author. The next stage of our demonstration will explore how intellectual property rights, recognizing an inventive intellectual work, can generate intangible capital.

7.3.2. *Intellectual works and intangible capital*

In this section, we will continue our analysis of Figure 7.5 in order to address the field of capital. We will begin by questioning the confusing profusion of epithets assigned to capital in the literature: intangible, intellectual and immaterial. Having set our sights on the intangible, we will successively shed light on intangible property, intangible assets and, lastly, intangible capital.

7.3.2.1. *Intellectual, immaterial and intangible capital*

7.3.2.1.1. Intellectual capital

"*Intellectual capital is intellectual property in action*" [KEE 99].

The idea of intellectual capital has proved to vary depending on the author [BIN 07]. Nicolas Binctin noted that the expression first appeared in the magazine *Fortune* in a paper by Stewart [STE 91]. The concept of intellectual capital has led to several publications for corporate managers [EDV 97, BON 98, STE 97, ROO 98], as well as academic publications [BON 99]. If these sources show a very heterogeneous nature of the elements grouped together by management sciences within the concept of intellectual capital, then they also allow us to observe that three elements are systematically addressed by authors:

– human capital: it cannot be appropriated by the organization as it combines knowledge, skills and the talents of individuals;

– structural capital: it brings together the knowledge outside human capital (internal organization, records of procedures, etc.) which cannot be appropriated (with the exception of specific elements, such as databases);

– intellectual properties: they stimulate appropriation mechanisms according to several systems (industrial property, copyright, etc.).

According to Nicolas Binctin, managers define intellectual capital as the company's entire intangible possibilities that are not registered and are not specifically valued by accounting records. Nonetheless, by keeping only the elements that can be appropriated, the intellectual capital, as seen by the management sciences, is reduced to the creations appropriated by one of the intellectual property regimes (primarily industrial property and copyright) or by secret [BIN 07].

7.3.2.1.2. Intangible capital

Intangible capital is a concept commonly used to refer to all intangible elements that can give the companies possessing them an economic advantage on the market [LAP 01a, BOU 06a, BOU 06b]. These elements can therefore be interpreted as intangible assets, according to Breesé and Kaiser [BRE 04a]. Consequently, logic would dictate that this capital be called intangible capital [SAU 13], which would be our choice of phrase, as indicated below.

Michel Vivant and Jean-Michel Bruguière note that "intellectual property occupies a central position in the economy of the intangible. Indeed, in an economy where ideas prevail and where it is innovation that creates value, it is normal that actors seek to protect these ideas or, at the very least, the economic benefit that they can derive from them" [VIV 09]. According to these authors, copyright is "the author's right on the basis that the work is the continuation of the individual". It follows that the author occupies a pre-eminent position and that he is granted moral rights to ensure respect for both the author as a person and for his work, as they are one and the same.

For Blandine Laperche, the notion of intangible capital is currently used to refer to all intangible elements whose possession is likely to provide a company with an economic advantage on the market [LAP 01a]. Blandine Laperche thus distinguishes between three different aspects of intangible capital in an innovative company:

– technological assets;

– elements related to commercial techniques and to methods of accessing customers;

– the organization and the individuals that form it. Some authors, incidentally, are quick to refer to an "organizational capital" [LAM 97].

7.3.2.1.3. Intangible capital

Our intention here is to shed light on use of the term "intangible" with regard to assets (intangible asset), and then capital (intangible capital).

Intellectual works create intellectual property rights, which form part of intangible assets. According to Patrick Ochs, an intangible asset is a dynamic, dematerialized investment that sustainably incorporates a part of the dominant knowledge in order to contribute, in a specific or procedural manner, to the competitivity and value of the company [OCH 96]. Breesé and Kaiser define intangible assets as follows: "The term 'intangible asset' refers to an asset that derives its value from an intangible existence, that is to say that it cannot be touched or seen, in contrast to a tangible asset that derives its value from its material existence and its physical properties" [BRE 04a]. These authors note that the classification of intangible assets differs depending on the author. For example, F. Reilly and P. Schweihs propose the following classification [REI 01]: marketing assets (brands, logos, trading names), technological assets (patents, laboratory notebooks, know-how), artistic assets (literary works, drawings and models), software assets (databases, computer programs, integrated circuits), engineering-related assets (industrial design, know-how, secrets, industrial drawings and blueprints, technical documentation), customer assets (client lists, contracts, customer relationships), contract-related assets (contracts on supply, licensing, franchising, non-competition), human capital (trained and organized employees, employee contracts), official licenses (mining rights, water access rights, airlines) and goodwill (celebrity, fame). In this way, we are able to distinguish, on the one hand, formalized intangible assets arising from an intellectual work open to formally registered property rights (industrial property rights) or not (literary and artistic property rights as the author's moral and patrimonial rights), and, on the other hand, non-formalized, tacit intangible assets that cannot be appropriated, as such, by the company.

Based on this analysis of intangible assets, we can logically, but not exhaustively, define intangible capital as the entirety of intangible assets.

7.3.3. *Traces in the inventive intellectual corpus*

We have defined the inventive intellectual corpus using the intangible envelop of a corpus [BLO 32, DAU 71], including all of the content of works arising from creative activity, regardless of their material support. As mentioned earlier, the corpus is used here under its meaning of entirety (of texts, documents) as applied to content only (corpus: an assembly of persons brought together under the same rules, hence: compendium of laws and registrations, after *corpus juris*). The intellectual corpus to be passed on is not made up of material supports, but by their content; for a long time, the transfer of this intellectual patrimony has only been carried out orally, while the content was passed down from generation to generation. Today, this patrimony (which is growing incessantly) is far too extensive to be transferred orally.

In practice, we will define the traces of the inventive intellectual corpus as the marks left by the material supports transcribing the inventive activity, grouped under the tangible twin of the intangible envelop of the inventive intellectual corpus. Within the framework of the organization, we will define this tangible twin, albeit not exhaustively, as all invention patents, articles and communications, study reports, thesis or study reports, internal memos, white papers, presentations, training aids, etc., provided that they describe an inventive intellectual activity.

7.3.3.1. *Defining an invention patent*

This is an example of an industrial property right that confers upon its holder (physical or moral person) a monopoly of exploitation on a technical invention for a maximum duration of 20 years, as well as the right to prohibit unauthorized third parties from reproducing the invention, as defined in the patent claims. Rather than industrial property, it would be appropriate to refer to legal property.

7.3.3.2. *The concept of patentability*

A technical invention, that is, a technological innovation that gives rise to a technical solution to a given technical problem, is only patentable if it fulfills the following three cumulative criteria: industrial application, novelty and inventive activity.

It becomes apparent that the invention must be a creation (and not a discovery), pursuing a technical (rather than esthetic) aim and must be finalized (and not simply be an idea or theory). This invention therefore falls within the field of industrial property (and not literary or artistic property).

The concept of patentability is very powerful, as it requires at once both novelty and inventiveness.

7.3.3.3. *Information contained in an invention patent*

According to the French National Industrial Property Institute (*Institut National de la Propriété Industrielle* – INPI), there are 35 million invention patents worldwide that represent 80% of scientific and technical information and the OECD estimates that 70% of information contained in invention patents is only found in invention patents. The information is classified carefully in 70,000 classes and it benefits from two crucial specific qualities: a standardized drafting plan adapted to innovation and links that facilitate the traceability of innovations [BRE 04b]. According to their characteristics, patent bases are the largest global knowledge management network.

7.3.4. *Synthesis of the inventive intellectual corpus*

We have seen that, starting as an intellectual work, the outcome of creative activity, industrial property generates intangible capital by creating a value that allows the dematerialized work to be associated with a right that could be appropriated, having the nature of an asset and which this double title comes to consider an object of trade. Intellectual property, with its "*droit d'auteur*", consecrates the creator as an author (of whom the work is considered to be the continuation) and validates the new knowledge that will go on to enrich the intangible envelop of the intellectual corpus forming the inventive intellectual corpus.

7.4. Analysis of the links between invention, innovation and the inventive intellectual heritage

7.4.1. *Links between industrial property rights and innovation*

Our aim here is to show the part played in innovation by information protected by industrial property rights.

Blandine Laperche noted that the invention patent, of which there were almost four million active titles worldwide in 2000 (with 83% filed in the United States, Europe and Japan), has become a powerful means of encouraging invention due to the recognition of the role of advanced technologies in stimulating economic growth [LAP 01b]; the exclusive monopoly on temporary exploitation awarded by the property title becomes the twin of innovation, or, according to Joseph Schumpeter's definition, of "the commercial exploitation of innovation" [SCH 47]. Indeed, without legal protection, the new technique or the new process may be recovered and copied when distributed on the market. The granting of a monopoly on temporary exploitation allows the rights holder to be able to recover the capital they invested in order to create the invention, prohibiting less creative competitors from effortlessly taking these innovations, as well as acting as intangible assets, contributes to both its financial development and the development of its image [BRE 04b]. It is therefore a matter of organizing the relationships between economic actors (franchises, patent licensing contracts, etc.) and protecting their rights in accordance with the rights of others (verifying freedom of exploitation). Blandine Laperche thus noted that "large companies are currently looking to protect as well as possible their innovation potential, which is made up of the information, knowledge, know-how, techniques and goods that have been formed or are in the process of being formed" [LAP 01b].

At a time when everyone agrees that the competitiveness of companies currently depends mainly on their ability to incorporate new knowledge into value-creating products and services, the strategic role of invention patents is difficult to question [COR 07]. Nevertheless, explicit knowledge, subject to the legal protection provided by intellectual property rights, is only part of the knowledge being mobilized in the competitive struggle [BAU 96].

In summary, the link between industrial property and innovation is largely supported by the link to the innovation process and is characterized by the fact that the protective role of industrial property is a powerful encouragement for innovation due to its right to a monopoly over exploitation.

7.4.2. *Links between industrial property rights and invention*

Our aim here is to show the part played by information protected by industrial property rights in the capacity for invention, whether directly (informational support) or indirectly (methodological support).

7.4.2.1. *Direct support of invention patents*

Published patents help researchers to explore new fields (knowledge of what exists and of what is protected, understanding of past stages and humility in the face of the future, descriptions of prior art and the traceability of technical design choices); patents and publications complement each other. Publications focus on explaining phenomena while patents are positioned further on in the process, on results, applications and the means used for these applications [BRE 04b]. Invention patents have acted as the innovation memory for two centuries [BRE 04b].

As a counterpart, the patent also has to meet its obligation to disseminate the scientific and technical content of the invention, allowing "skilled individuals" to be able to reproduce it. The fact that the patent does not protect the idea (or the knowledge), but its materialization ensures compatibility between private incentives for invention (by increasing the personal yield from the invention) and technical progress (by disseminating knowledge).

A patent describes prior art and explains how the proposed solution differs from the existing one, thereby allowing an excellent level of traceability in the choice of technical designs [BRE 04b].

Since the industrial revolution, patents and technical standards have been combined to promote incentives for invention [LAP 01b].

It is possible to generalize the statements about invention patents to technical information and knowledge. "Information, knowledge and Industrial Property form a whole" [BRE 04b]. The economic perspective provides an interesting point of view distinguishing between knowledge and information; the knowledge has something greater than information as it "gives its holder the ability to take intellectual or physical action" [FOR 00]. "Knowledge is that which gives humankind cognitive abilities. Information, on the other hand, remains a set of data, doubtlessly formatted and structured, but passive and inactive by being unable to grant its bearer the capacity to take action" [FOR 00].

7.4.2.2. *Indirect methodological support for invention patents*

Formalizing an innovation "like a patent" [KER 01], that is, by describing certain aspects with documents that follow the editorial rules of patents, can contribute to strengthening this common vision of the work being created (what field is in the invention in? What is the problem to be solved? How is the problem currently solved?), in such a way that, by appropriating the specific style of invention patents, the team behind the innovative project:

– clearly defines its objectives;

– finds out the state-of-the-art;

– can imagine a wide variety of solutions;

– can compare its solutions to others in a synthetic way;

– strictly formalizes what it wants to do;

– communicates more effectively;

– is more aware that it is teamwork that allows "the best solution" [KER 01] to be created.

Formalizing and creating are therefore indivisible [KER 01]; "patent style" formalization is particularly suited to the inventive approach. Pierre Breesé and Yann de Kermadec confirm that reinvention is highly expensive and a record of the company's creations should be kept. Using an innovation base (that is, a knowledge base dedicated to innovation products and processes) allows ideas to be saved so that they can be used at the right time, as well as avoiding reinvention, and identifying problems, solutions and the people to contact. It contains innovation files, innovation memos, patents, files of inventions that have not been retained, etc. [BRE 04b].

In this way, the link between industrial property and innovation is shown in the fact that databases of invention patents are unrivaled and often exclusive sources of technological and scientific information, presented according to powerful formalism, methodologically speaking.

7.4.3. *Links between invention and industrial property rights*

Intuitively, it is easy to see that inventive activity fosters invention patents. Moreover, the promotion of research, which is part of the task of researchers, is carried out by granting patent licenses on the basis that different kinds of partnerships can be made with industrial companies [BRE 04b].

7.4.4. *Links between innovation and industrial property rights*

In this case, intuitively, it is also easy to see that innovation, as a product that has seduced the market, feeds brands, and also designs and models. Brands allow a product or service to be distinguished from other products and services, while designs and models allow highly varied kinds of esthetic creations to be protected [BRE 04b].

7.4.5. *Links between invention and innovation*

We have seen that the generation of ideas, the main component of inventive activity, falls within what the left part of Figure 7.2 showed as the "Front End of innovation", for a commercial approach to innovation. This is therefore the part that precedes, and fosters, the innovation process.

7.4.6. *Links between innovation and invention*

From our perspective, the product of innovation cannot directly foster inventive activity, precisely because it comes after the innovation process and is therefore the result of new knowledge that has been implemented [SAU 13].

7.4.7. *Reciprocal links between inventive activity and the inventive intellectual corpus*

Our *ICAROS*® method uses analysis of the inventive intellectual corpus to stimulate inventive activity [SAU 16a]; the inventive intellectual corpus of an organization contains basic elements that allows for the representation of a certain number of inventive trajectories that have been undertaken by the organization's knowledge actors by linking the inventive intellectual traces that have been left (patents, theses, reports of advanced studies). A reasoned analysis of the traces thus restores the intellectual path drawn by the trajectories of these traces (method acquisition, in-depth studies, investigation of variants, bifurcation, abandonment, etc.); we thus obtain a strategic evolution, in terms of the firm, of the inventive intellectual corpus that has been collected and stored over several years. The assessment obtained by this process of collecting and explaining inventive knowledge that has led to a new design is then put into perspective with the potential knowledge in each field (representing an absolute reference) to form a cognitive stimulus applied to experts representing each area of knowledge, who individually have the potential for the identification and the explanation of which they have been guided to in order to help them develop a forward-looking vision of their field [SAU 16a].

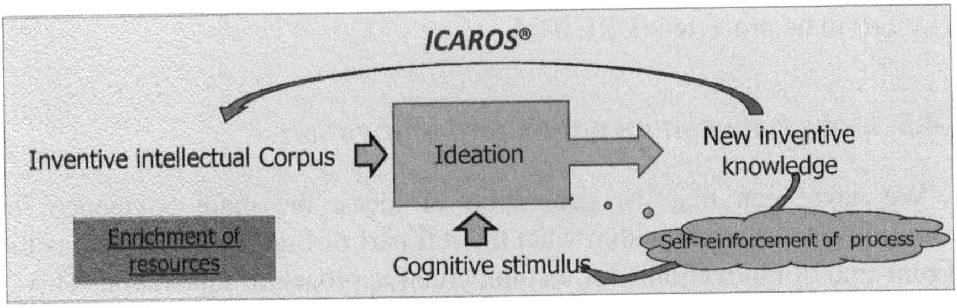

Figure 7.6. *Diagram of the production of inventive knowledge by ideation (source: Pierre Saulais)*

The reciprocity is expressed through the analysis, extraction and implementation of the inventive knowledge, with this new inventive knowledge going on to enrich the inventive intellectual corpus [SAU 16b, SAU 16c], as shown in Figure 7.7, which focuses on the dossier on inventive activity called the Invention and Innovation Dossier.

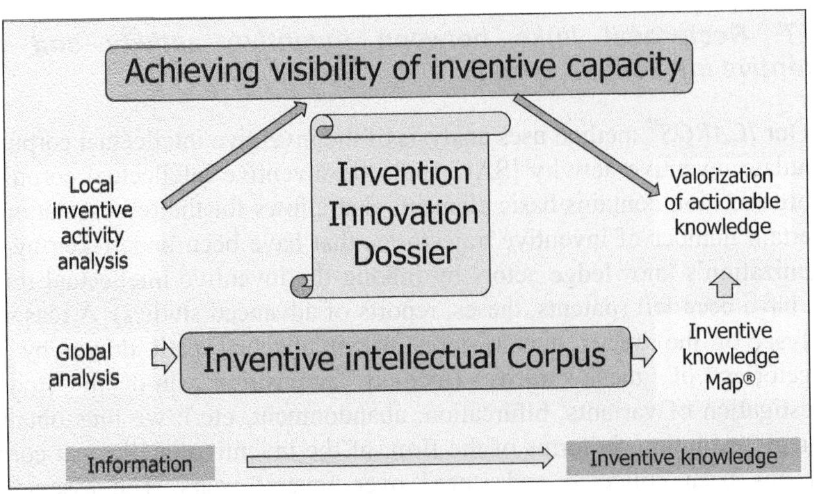

Figure 7.7. *Analysis of inventive activity (source: Pierre Saulais)*

Figure 7.8 shows the applications of analysis of overall inventive activity (contained within the inventive intellectual corpus) to support decision-making in different fields.

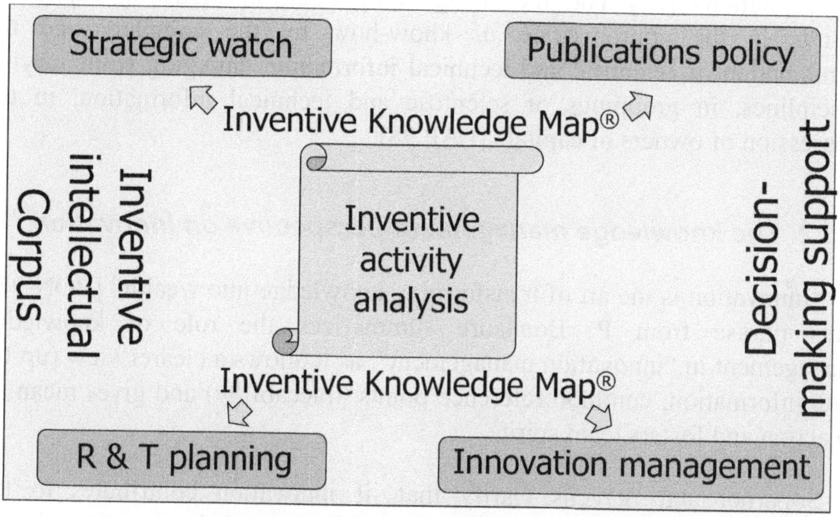

Figure 7.8. *Applications of showing analysis of inventive activity in knowledge maps (source: Pierre Saulais)*

7.5. The nature of bridges between knowledge areas

Our aim here is to show that the links mentioned above between two components, where one fosters the other, have a much deeper nature, which is one of the bridges between several areas of knowledge. The knowledge objects that we have explored are now considered to be representative of a field of knowledge. We can now question the nature of these bridges.

7.5.1. *The perspective of economists*

The perspective of economists focuses their frame of reference on organization: "Understanding competition, analyzing risk and strategies for appropriation and trade means understanding why innovation and technical change result from learning and the accumulation of scientific, technical and productive information and knowledge over long periods of time. Based on this situation, the information production systems are definitely part of industrial patrimony and the productive capital of the company that finances and/or appropriates them" [UZU 98]. "Information, especially that arising from scientific and technical works, is a commodity […] it is a capital asset that accrues, in order to be enhanced, over the development and expansion of trade relations […] The increasing competition in industrial capitalism results in the appropriation of know-how by the company and the combination of scientific and technical information emerging from varying disciplines, in groupings of scientific and technical information, in the possession of owners of capital" [LAP 98].

7.5.2. *The knowledge management perspective on innovation*

"Innovation is the art of transforming knowledge into wealth" [BON 97]. This phrase from P. Bonnaure summarizes the role of knowledge management in "innovation management", as it allows a clearer view (up-to-date information, common reference points, traceability) and gives meaning to action and fosters team spirit.

Soparnot and Stevens clarify that, if innovation contributes to the regeneration of the organization by stimulating knowledge creation, then it also equally depends on it, as it is only able to develop thanks to the organization's accumulated knowledge; innovation management therefore invites knowledge management as, in part, knowledge determines innovation just as it comes from it [SOP 07].

7.5.3. *The perspective of KBI (Knowledge-Based Innovation)*

This is the first level of application for our *ICAROS®* method [SAU 16a]. This level is defined by the fact that the nature of bridges between the three fields of knowledge is characteristic of a flow of knowledge, which will allow each field to benefit from the others (see section 7.4.7). The analysis, mapping and implementation of the inventive knowledge are a definite part of the rise of the knowledge value chain, using metacognition and control over decision-making in an ever more complex world [SAU 16c]. Such an approach, which is still to be instrumented, would allow a company, seen as an information processor, to make all aspects of its strategic activity consistent (see Figure 7.8). This approach, developed within the KBI framework, paves the way for IKM (Inventive Knowledge Management).

7.5.4. *The perspective of knowledge-based ideation*

This is the second level of applying our *ICAROS®* method [SAU 16b, SAU 16c], characterized by a greater degree of abstraction (and therefore more specifically applicable to the two knowledge fields of inventive activity and the inventive intellectual corpus). This abstraction allows us to consider that the ideation (in the sense of generating inventive ideas), central to inventive activity, is of the same nature as its outcome (the inventive idea, considered to be new inventive knowledge), which will itself enrich the inventive intellectual corpus, whose analysis and extraction of inventive activity (of the same nature as the outcome they produce) will themselves also stimulate ideation by their outcome (see Figure 7.6); it is therefore a phenomenon of survenience. Moreover, when the inventive idea is put into action in an intellectual work recognized as such by the granting of intellectual property rights, the inventor is recognized as an author (in the sense of "*droit de l'auteur*"), who is regarded as one with his work, since the author makes the work, just as much as the work makes the author. In this way, the nature of the bridge between inventive activity and inventive intellectual corpus (having given rise to the granting of intellectual property rights) is therefore the knowledge of knowledge (or metacognition):

"It is the faculty of knowing that organizes knowledge and not the objects that determine it. The center of knowledge is the knowing subject: how does our knowledge reflect our cognition?" [KAN 81].

7.6. Conclusion

This exploratory study has allowed us to define invention, innovation and intellectual property rights as objects of knowledge, with the aim of exploring their nature and then their genetic interactions.

We have thus presented each of the three objects of knowledge, primarily the act of innovating and its outcome, followed by the ability to invent, and lastly the inventive intellectual corpus and intangible capital. By considering each object of knowledge to represent its own field of knowledge, we have examined the links between the three fields both in a static sense and in a dynamic sense of one thriving on the other. In our examination of the nature of these links, we responded by characterizing our *ICAROS®* method in two levels of abstraction; according to the first level, the bridges between fields have the nature of inventive knowledge flows moving from one field to another. According to the second level, it is a matter of the flow generated by the knowledge of this knowledge; that which fosters inventive knowledge in a field (and allows the creation of new inventive knowledge there) is therefore analysis of the structure of the knowledge deposited in the inventive knowledge in another field.

7.7. Bibliography

[ALL 03] ALLAND D., RIALS S., *Dictionnaire de la culture juridique*, Presses Universitaires de France, Paris, 2003.

[ALT 84] ALTSHULLER G., *Creativity as an Exact Science: The Theory of the Solution of Inventive Problem*, Gordon and Breach Science Publishing, New York, 1984.

[ALT 98] ALTSHULLER G., *The Innovation Algorithm: TRIZ, Systematic Innovation and Technical Creativity*, Technical Innovation, 1998.

[ALT 00] ALTER N., *L'innovation ordinaire*, Eyrolles, Paris, 2000.

[AMI 01] AMIDON D., *Innovation et management des connaissances*, Editions d'Organisation, Paris, 2001.

[ARG 93] ARGYRIS C., *Savoir pour agir*, Jossey-Bass, San Francisco, 1993.

[BAT 72] BATESON G., *Steps to an Ecology of Mind: Collected essays in Anthropology, Psychiatry, Evolution and Epistemology*, University of Chicago Press, Chicago, 1972.

[BAU 96] BAUMARD P., *Organisations déconcertées*, Masson, Paris, 1996.

[BEN 08] BENOIT-CERVANTES G., *La Boîte à outils de l'Innovation*, Dunod, Paris, 2008.

[BIA 74] BIANCHI H., *L'innovation et ses contraintes: études de cas*, Eyrolles, Paris, 1974.

[BIN 07] BINCTIN N., *Le Capital Intellectuel*, Litec, Paris, 2007.

[BLO 32] BLOCH O., VON WARTBURG W., *Dictionnaire étymologique de la langue française*, Presses Universitaires de France, Paris, 1932.

[BON 97] BONNAURE P., "Les politiques d'innovation", *Futuribles*, no. 225, 1997.

[BON 98] BONTIS N., "Intellectual capital: an exploratory study that develops measures and models", *Management Decision*, vol. 36, no. 2, pp. 63–76, 1998.

[BON 99] BONTIS N., "Managing organizational knowledge by diagnosing intellectual capital: framing and advancing the state of the field", *International Journal of Technology Management*, vol. 18, nos. 5–8, pp. 433–462, 1999.

[BOU 01] BOUTILLIER S., "L'entrepreneur face à la propriété industrielle: histoire et actualité", in LAPERCHE B. (ed.), *Propriété industrielle et innovation*, L'Harmattan, Paris, 2001.

[BOU 06a] BOUNFOUR A., *Capital immatériel, connaissance et performance*, L'Harmattan, Paris, 2006.

[BOU 06b] BOUNFOUR A., EPINETTE G., *Valeur et performance des Systèmes d'Information*, Dunod, Paris, 2006.

[BRE 04a] BREESÉ P., KAISER A., *L'évaluation des droits de propriété industrielle*, Gualino, Paris, 2004.

[BRE 04b] BREESÉ P., KERMADEC Y. DE, *La propriété intellectuelle au service de l'innovation*, Nathan, Paris, 2004.

[CAL 85] CALLON M., LATOUR B., "Comment suivre les innovations, clefs pour l'analyse socio-économique", *Prospective et santé publique*, no. 10, 1985.

[CAV 07] CAVALUCCI D., ELTZER T., "Parameter network as a mean for driving problem solving process", *International Journal of Computer Applications in Technology*, vol. 30, no. 1, pp. 125–136, 2007.

[CHA 99] CHANAL V., "Management de l'innovation de produit industriel : mise en oeuvre d'une démarche de diagnostic pour améliorer notre compréhension du processus", *7ème conférence de management stratégique*, Ecole Centrale de Châtenay, 1999.

[COR 87] CORNU G., *Vocabulaire juridique*, Presses Universitaires de France, Paris, 1987.

[COR 07] CORBEL P., *Management stratégique des droits de la propriété intellectuelle*, Gualino, Paris, 2007.

[DAU 71] DAUZAT A., *Nouveau dictionaire étymologique et historique*, Larousse, Paris, 1971.

[DEC 80] DECLERCK R.-P., EYMERY P., CRENER M.-A., *Le management stratégique des projets*, Editions Hommes et Techniques, Paris, 1980.

[EDV 97] EDVISSON L., MALONE M., *Intellectual Capital: Realizing Your Company's True Value by Finding its Hidden Brainpower*, Harper Business, New York, 1997.

[FER 06] FERNEZ-WALCH S., ROMON F., *Management de l'innovation*, Vuibert, Paris, 2006.

[FOR 00] FORAY D., *L'économie de la connaissance*, La Découverte, Paris, 2000.

[GAF 34] GAFFIOT F., *Dictionnaire Latin-Français*, Hachette, Paris, 1934.

[GAR 95] GAREL G., MIDLER C., "Concourance, processus cognitifs et régulation économique", *Revue française de gestion*, no. 104, pp. 86–101, 1995.

[GAR 02] GARCIA R., CALANTONE R., "A critical look at technological innovation topology and innovativeness terminology: a literature review", *Journal of Product Innovation Management*, vol. 19, pp. 110–132, 2002.

[GIA 93] GIARD V., MIDLER C., *Pilotage de projets et entreprises. Diversités et convergences*, Economica, Paris, 1993.

[GUE 09] GUELLEC D., *Economie de l'Innovation*, La Découverte, Paris, 2009.

[HAT 94] HATCHUEL A., "Apprentissages collectifs et activités de conception", *Revue française de gestion*, no. 99, pp. 109–120, 1994.

[HAT 02] HATCHUEL A., WEIL B., "C-K theory: notions and applications of a unified design theory", *Herbert Simon International Conference on Design Sciences*, Lyon, 2002.

[HEN 90] HENDERSON R., CLARK K., "Architectural innovation: the reconfiguration of existing product technologies and the failure of established firms", *Administrative Science Quaterly*, vol. 35, no. 1, pp. 9–30, 1990.

[KAN 81] KANT E., *Critique of Pure Reason*, 1781.

[KEE 99] KEEN P.G., *Transforming Intellectual Property into Intellectual Capital: Competing in the Trust Economy*, Stanford, Hoover Institution Press, 1999.

[KER 01] KERMADEC Y. DE, *Innover grâce au brevet*, Dunod, Paris, 2001.

[LAM 86] LAMBIN J.-J., *Marketing stratégique*, McGraw Hill, New York, 1986.

[LAM 97] LAMBERTON D., "Pour une taxinomie de l'information", in MAYERE A. (ed.), *La Société informationnelle*, L'Harmattan, Paris, 1997.

[LAP 98] LAPERCHE B., *La firme et l'information: innover pour conquérir*, L'Harmattan, Paris, 1998.

[LAP 01a] LAPERCHE B., *Propriété intellectuelle et Innovation*, L'Harmattan, Paris, 2001.

[LAP 01b] LAPERCHE B., "Brevets et normes techniques: de l'incitation à l'invention au contrôle de l'innovation", in LAPERCHE B. (ed.), *Propriété industrielle et Innovation*, L'Harmattan, Paris, 2001.

[LE 09] LE LOARNE S., BIANCO S., *Le management de l'innovation*, Pearson Education, 2009.

[LIT 86] LITTRE E., *Dictionnaire de la Langue Française*, Hachette, Paris, 1886.

[LOU 08] LOUAFA T., PERRET F.-L., *Créativité et innovation*, Presses Polytechniques, Lausanne, 2008.

[MAF 98] MAFFIN D., "Engineering design models: context, theory and practice", *Journal of Engineering Design*, vol. 9, no. 4, pp. 315–327, 1998.

[MAR 91] MARCH J., "Exploration and exploitation in organizational learning", *Organization Science,* vol. 2, no. 1, pp. 71–87, 1991.

[NAV 89] NAVARRE C., SCHAAN J.-L., "Où en est la gestion de projet aujourd'hui? Essai de modélisation", *Gestion 2000*, vol. 5, no. 1, pp. 31–57, 1989.

[NON 97] NONAKA I., HIROTAKA T., *La Connaissance créatrice*, De Boeck, Bruxelles, 1997.

[OCH 96] OCHS P., "Investisement immatériel et indicateurs de pilotage", *Supplément à la revue Echange*, no. 119, pp. 60–66, 1996.

[PRA 05] PRAX J.-Y., BUISSON B., SILBERZAHN P., *Objectif Innovation: Stratégies pour construire l'entreprise innovante*, Dunod, Paris, 2005.

[REI 01] REILLY R., SCHWEIHS R., *Valuing Intangible Assets*, Irwin Library – McGraw Hill, 2001.

[ROO 98] ROOS J., ROOS G., EDVISSON L. *et al.*, *Intellectual Capital: Navigating in the New Business Landscape*, New York University Press, New York, 1998.

[ROU 73] ROUQUETTE M.-L., *La Créativité*, Presses Universitaires de France, Paris, 1973.

[SAU 13] SAULAIS P., Application de la gestion des connaissances à la créativité des experts et à la planification de la R&T en milieu industriel de haute technologie, PhD Thesis, Télécom Ecole de Management, 2013.

[SAU 16a] SAULAIS P., ERMINE J.-L., "Emergence d'innovation technologique incrémentale par application, à des experts, de l'analyse du patrimoine intelectuel inventif", *Innovations*, vol. 49, no. 1, pp. 103–140, 2016.

[SAU 16b] SAULAIS P., "Mise en visibilité du potentiel inventif et représentation du patrimoine intellectuel inventif", *Veille Mag*, no. 155, 2016.

[SAU 16c] SAULAIS P., LECANTE C., "Analyse, cartographie et mise en action de la connaissance inventive", *Actes du 9ème Colloque International GeCSO*, Paris, 2016.

[SCH 47] SCHUMPETER J., *Capitalisme, Socialisme et Démocratie*, Payot, Paris, 1947.

[SEN 91] SENGE P., *La cinquième discipline*, Editions Générales First, Paris, 1991.

[SIM 69] SIMON H., *The Science of the Artificial*, MIT Press, Cambridge, 1969.

[SOP 07] SOPARNOT R., STEVENS E., *Management de l'innovation*, Dunod, Paris, 2007.

[STE 91] STEWART T., "BRAINPOWER Intellectual capital is becoming corporate America's most valuable asset and can be its sharpest competitive weapon. The challenge is to find what you have – and use it.", *Fortune magazine,* vol. 123, no. 11, p. 44, 1991.

[STE 97] STEWART T., *Intellectual Capital: The New Wealth of Organizations*, Doubleday, New York, 1997.

[TID 06] TIDD J., BESSANT J., PAVITT K., *Management de l'innovation: intégration du changement technologique, commercial et organisationnel*, De Boeck, Bruxelles, 2006.

[UZU 98] UZUNIDIS D., "Préface", in LAPERCHE B. (ed.) *La Firme et l'Information: Innover pour conquérir*, L'Harmattan, Paris, 1998.

[UZU 01] UZUNIDIS D., "La marchéisation de la recherche publique en France et la question de la propriété industrielle", in LAPERCHE B. (ed.) *Propriété industrielle et Innovation*, L'Harmattan, Paris, 2001.

[VIV 09] VIVANT M., BRUGUIERE J.-M., *Droit d'auteur*, Dalloz, Paris, 2009.

[WAT 75] WATZLAWICK P., FISCH J., WEAKLAND R., *Changements: Paradoxes et psychothérapies*, Seuil, Paris, 1975.

8

Commitment to an Industrial Evolution

If the term "industrial revolution" rings a bell (we are already on the verge of the fourth one[1], and have been since the first reference was made to the arrival of "smart factories" at the CEBiT expo in Hannover in 2010 [CEB 16]), the idea of "evolution" strains the brain, serving only to confuse.

There are three major potential causes for this: first, evolutions are rare gems. According to Ken Wilber and taken up by Frédéric Laloux, we would be experiencing a fifth major transformation in the history of humankind, putting into perspective the four industrial revolutions that have taken place over barely two centuries [LAL 14]. In this way, after having successively plundered the magical/tribal, traditional/agricultural, scientific/industrial and post-modern/computer science ages, we would be experiencing a profound holistic transformation. Jean Staune makes clear the five *re*volutions related to this transformation [STA 15, STA 16]: technological, conceptual, economic, societal and managerial. On the basis of these postulates, we would be able to question the way in which industry integrates (itself into) this overall vision, the ins and outs of which can no longer be reduced to a mere productionist framework, strictly speaking.

The second cause of perplexity regarding "evolution" lies in the extreme complexity that the term implies in modern society, which has multiple and wonderfully interconnecting impacts, and at the forefront of which are

Chapter written by Theodor FELEZEU.

1 The four revolutions can be identified as follows: 1st revolution (19th Century) – machines powered hydraulically (steam), 2nd (beginning of the 20th Century) – electrical energy and mass production (Taylorism), 3rd (1970s) – electronics and information tecToyotism) and 4th (2010) – Internet of Things, Big Data, additive and modular manufacturing.

employment and the environment, which remain its cornerstones. Even there, what could the responses of industrial renewal be to these major political challenges?

Lastly, while we have a certain affinity with revolutions, whether to overturn existing powers or to market the latest generation iPhone, there are many who, when daring to address evolution, hasten to call upon Darwinism. Without necessarily indulging in such reductive talk, or even, on the contrary, redefining the concept itself, we have (and the industrial world *a minima* has) the duty to return meaning to the term in the present context, to treat it with the high standards required by the complexity of the world we live in today, and in which our children and grandchildren will play tomorrow. We can therefore wonder how to address industrial inventions and innovations so as to provide this collective requirement (our own) with a durable and sustainable dynamic.

As such, after having made observations and pointed out the risks incurred by current industrial policies, this chapter will explore three direct lines on which industry must rely in order to reinvent itself along a truly sustainable line: industry X.0, the values at the heart of production systems and the emergence of new economic models. The end of the chapter will illuminate matters with corollaries and a summary devoted to the impact of this industrial evolution on cultures, and the visions, strategies, processes and tools of innovation.

8.1. Findings, shared feelings and risks incurred

In order to examine certain global phenomena through an industrial lens, it is interesting to consider some data related to the economic, social and environmental effects caused by industrial policies currently being undertaken:

– the share of industry in the global economy rose, in 2014, to 27.6% [BAN 16];

– 3.1 million employees worked in industry in France in 2014, excluding temps [GOU 16];

– the advent of industry 4.0 involved the loss of 5 million jobs [LAT 16];

– 1/2000 will be the yield of rare earths/toxic waste [KAI 14] involved in the exploitation, to date, of 17 so-called "strategic" metals used particularly in the production of touch screens, contact lenses, wind turbines, low energy light bulbs and even car batteries;

– 1.36/2.25/1.18, the multiplying coefficients of greenhouse gas concentrations since 1850 for CO_2/CH_4/N_2O, according to the French Group of Intergovernmental Experts on Climate Change (Groupe d'Experts Intergouvernemental sur l'Evolution du Climat – GIEC) at the beginning of a climatic change that has now been proven [IPC 14].

Figure 8.1. *Homo numeris*

In the current state of affairs, it seems difficult to be happy about these findings or the ensuing prospects, and especially in the event that concrete measures, affecting different spatio (societies, territories, nations/economic powers, consumers)-temporal (very short- and medium-term) scales, would not be undertaken with sufficiently demanding requirements, i.e. enabling an enduring harmony between Man and its environment(s).

In this way, if we fail to meet these strategic challenges, that represent the suffixes "-nomy" and "-logy" that follow "eco" (a number of intellectuals including Jacques Attali, Christian Saint-Etienne and even Philippe Baumard [BAU 12, BAU 14] have highlighted a preponderance of tactical action and short-termism at the expense of consideration of long-term impacts), there is a high risk of being exposed to a "Game Over" scenario, that our societies continue to court and even increase by forging the sword of Damocles that hangs over us, and the future of our children and grandchildren …

… unless we are still able to continue to pursue a mythological narrative of sacrifice, maintained (consciously or not) by a dominant economic model that is profoundly unequal, not only in terms of the present generations (in 2016, 1% of people in the world had more than the other 99% [OXF 15]), but also future ones (progressive acceleration of the debt economy since the 1970s [INS 16]).

In other words, each time that decisions are made on the basis of research into maximum profitability, most often leaving in its wake theories of unbridled competitiveness and blind growth (an increasingly ill-treated myth given the current paradigm linking demographics – and therefore trends in consumption – and production – to date, waste generation in particular), there is a substantial risk of relegating human lives and individuals' trajectories to the background. The consequences of this risk are indicated by a number of individuals in the intellectual world, such as the philosopher Jean-François Mattéi in his most recent work [MAT 15] or the lawyer Alain Soupiot, a French college professor [SUP 15].

For example, the disproportionate (thanks to a (hyper-) financial mechanism) subtraction of capital from production facilities (taking into account the challenges they face in terms of social and environmental utility) using a shareholding paradigm based on profit maximization, which began in the 1970s only to blossom in the mid-1980s (see Figure 8.2).

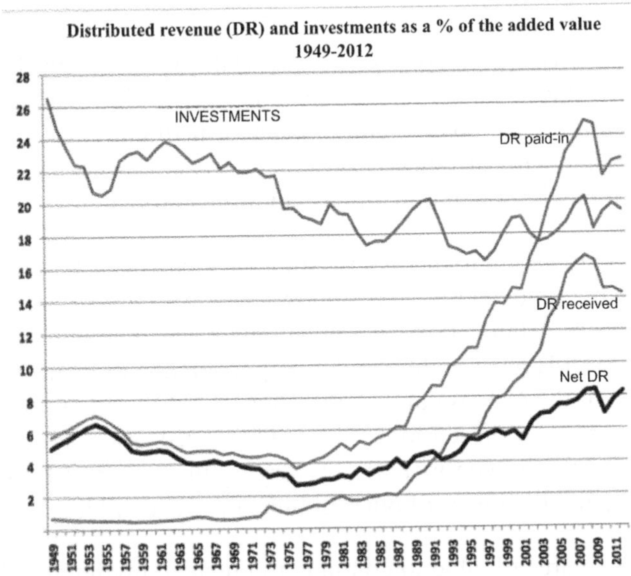

Figure 8.2. *Revenue distributed to shareholders of non-financial corporations (NFC) in France in comparison to the investments made [GAD 14] (source: Jean Gadrey)*

Jean Gadrey, an honorary professor of economics at Lille 1 University explained that, "in effect, we can estimate that with 3–4% of the added value between the 1960s and the 1980s, before the financial revolution, the dividends and other net distributed revenue were therefore compatible with the effective production operations of businesses. As with 7.5–8% in recent years, we have around 4% unproductive shareholder income […] 40 billion Euros […]. Not only unproductive, but also destructive since the essential element goes into destabilizing speculations." Imagine the effect of distributing this 10–15 billion in a modernized industry with its own means of production and teams formed to combat the challenges of the future.

So, in concrete terms, what choices do we have? Let us consider four options:

immobilism, blindness – or loyalty [DEN 14]: the world moves forward on an unchanging trajectory, based on paradigms and habits arising from the "ancient world" (infinite growth, maintenance of inequalities) within the background of tactical makeshift efforts with the aid of control (generally illusory in the Internet age) of public opinion;

flight: yes, but towards what? Forward? (see previous option);

rebellion: dear to indecisive, anti-establishment revolutionaries, this includes an (often) violent destruction of previous models, glorifying primal instincts and cultural isolationism, civilizational conflicts and divisions, bringing about the risk of a chain reaction of losing points of reference and of social, societal and inter-generational splits;

relevance: based on a dual awareness/move to action, tested by the creative and engaged elites challenging the ill-suited paradigms of the past, progressing in concentric circles (from the local level to the global) to build the well-being of societies (human and environmental) in an efficient and durable manner for now and beyond. What remains (all things being equal) is to transform the good old project and its V-model (study, specification, design, validation, delivery) into an agile "Lean" model (prototyping, testing/analysis, implementation), as perfectly described by Ries [RIE 11].

Presented like this, it is enough to understand the present in order to have an overview of the ambient noise whose (increasingly strong) signals appear in the economic, environmental, political, religious or identity-based crises that we are experiencing. Depending on the influences exercised or endured, we are led to position ourselves in relation to these four choices, making us, in a pseudo-permanent, concurrent and *a fortiori* chaotic manner, into hostages to the tides (and not only migratory or financial ones) that appear more and more difficult to control. This is reminiscent of the unrest experienced by our ancestors in circumstances contextually related to political regime reversals, technological upheavals or even natural disasters, which even then were a sign of a profound change. The "appeal" of our era lies in the facts that all of these changes occur on a global scale and that we are given the opportunity (or the risk) to experiment on our own lives.

The first problem related to this unusual, even "superhuman" (which, by means of states and companies, the major players in these challenges *de facto* are), situation is that the fourth path seems clearly under-represented in relation not to the three other paths, but to the effect caused over time by the three others: the impact of the legacy passed down to us over 200 years of fierce exploitation, completely antagonistic ideologies, archaic forms of awareness of progress, mass education, heightening competitive logic, downward spirals, various abandonments, etc. As well as, lastly, choices and decisions considered appropriate at the time they were implemented (a time which has passed, and which continues, even as we gradually move towards an increasingly fast pace of life), which eventually settles into clear ideological grooves and habits, which, when the moment to bring about change arrives, are particularly difficult to change.

The second problem lies in the manner of addressing this fourth path, which many of us do not consider in a sufficiently comprehensive or holistic way, isolated within outdated arguments in other schools of thought (many thinkers in France in particular), generating endless debates on the way to think about the economy (market vs. solidarity), society (entrepreneurship vs. state subsidies), companies (giant vs. start-up), digital technology (platforms vs. blockchain), political activity (state/interventionist vs. citizen/libertarian), etc.

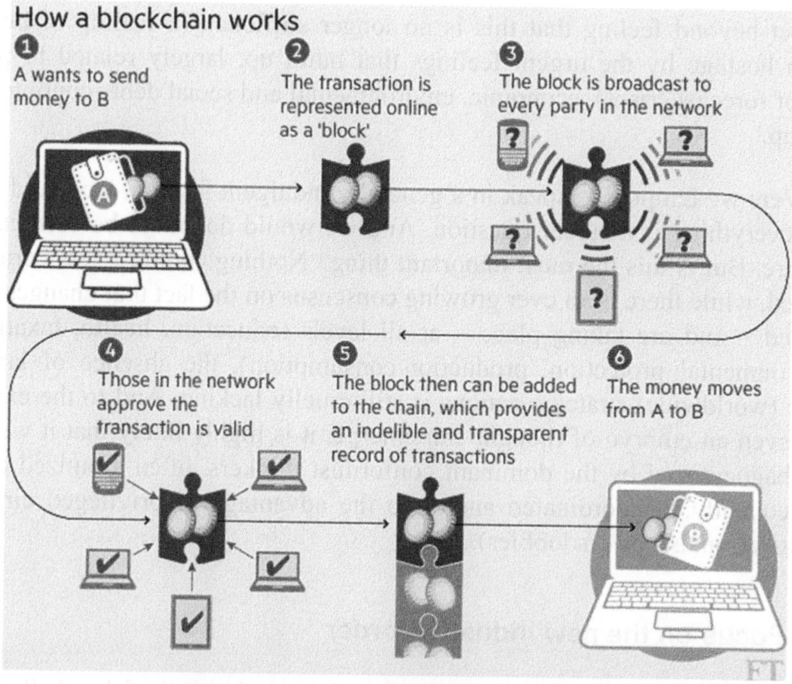

Figure 8.3. *Operating principles of blockchain, an exploratory tool (source: Financial Times)*

Just as Cartesianism[2], deeply rooted in values, had overcome the art of the paradox, it is useful to address this in a pluralist, complicated, even "quantum" universe.

Moreover, when this path is chosen and handled with primarily good intentions, the majority of the time it will doubtless have been praised from a less demanding context; less polluted, changing the unemployment trend, content with arrangements in education while real structural reforms are needed …

2 Through its two main components (determinism i.e. the "mechanical" link between cause and effect, and reductionism i.e. the systematic division of an issue into smaller tasks, each of which is itself divided until it is reduced to specific modules, each responding to a simplified problem), taking part in an "atomization" of roles and functions, e.g. "generalist"/ "specialist" divides or, more generally, in a worldview resulting in modes of thinking full of Manichaeism against which some "masters" of complex thought such as Jean Staune and Edgar Morin [MOR 11] intervene.

Yet beyond feeling that this is no longer sufficient (a feeling of being taken hostage by the urgent feelings that build up, largely related to past and/or forecast crises), economic, environmental and social debts continue to pile up.

Were we tempted to speak in a generally indulgent fashion, we could say that everything is a cursor question. And we would doubtless be correct, to be sure. But is this the most important thing? Nothing could be less certain. Indeed, while there is an ever growing consensus on the fact that changes are needed – and are taking place – at all levels (education, health, taxation, environmental protection, production/consumption), the absence of large-scale (worldwide) strategic actions is still cruelly lacking. And to the extent that even an embryo of thought can emerge, it is highly likely that it would be phagocytozed by the dominant conformist thinkers, often atomized and, consequently, uncoordinated and/or to the advantage of privileged circles (multinationals/powers/lobbies).

8.2. Focus on the new industrial order

And as for industry in all this? In our opinion, taking all of the challenges mentioned into consideration, it has a fundamental role to play through its ability to renew itself. It cannot be content with a mere revolution, regardless of the technological, operational, managerial or social and environmental "quantum leap" under the current system of corporate social responsibility. Needless to say, this would be a "commendable" step forward. However, it definitely remains insignificant in terms of impact in relation to a real "change in the rules of the game", such as the one extolled by industrialist Gunter Pauli, founder of the Blue Economy [PAU 15]. His vision? Not only to pollute "less", but quite simply not to pollute at all! The aim is production with ZERO WASTE.

It is at this stage that we arrive at the crux of the subject. There are questions we must ask. Transposed to the aforementioned findings, can industry free itself from the existing economic and political rules? Not entirely. With the advent of uberization and technological changes that are already well underway? With no advantage, how could industry do it? Does it nonetheless have the duty, the requirement to reinvent its very nature in order to respond to future challenges? Everything depends on the meaning that either politicians, economists, heads of industry or civil society wish to give this future. The meaning takes precedence over the cursor. It is a "yes" to innovation, but *with* and *for* mankind.

IMPORTANT.– Therefore, to sum up: the need for a collaborative and multidisciplinary reinvention based on fundamental and inalienable values, accompanying and fostering the creative experimentation of entrepreneurs.

Concretely speaking, this could rely on three ambitious projects, which, we should note, have the good sense to coexist in (still!) separate fields, which are far from cooperating given the requirements mentioned in the introduction:

industry X.0: digital, connected, collaborative and almost bespoke;

illuminated by Renaissance values: humanity, education and knowledge;

supported by alternative economics: rethinking revenue, labor value and sharing.

8.2.1. *Beyond industry 4.0*

Of the three projects, the one related to connected industry and to its new forms of production, seems to be *a priori* the most comprehensible. Various players in the world of consultancy in particular explain that these approaches, which have arisen from the confluence of technological progress and *lean* techniques developed by industry during the so-called Thirty Glorious Years in France (1945–1975), target more effective and more optimized production. Kurt Salmon therefore refers to the key principles of Industry 4.0 [SAL 15], which "constitute a true breaking point in three major technological innovations: collaborative robotics, self-adaptation of production systems (thanks to intelligent and connected sensors) and recourse to enhanced reality". Consequently, following Germany (2010), the United Kingdom (2011) and Italy (2012), France launched its own "Factory of the Future" program in 2013. Within the framework of a roadmap with 34 plans for a new industrial France, it aims to modernize industries – France suffers from an investment shortfall of 40 billion Euros in its industry in comparison to its German neighbor – and thereby increase competitivity and create jobs [GOU 14].

RENEWABLE ENERGY
A CAR FOR ALL CONSUMING LESS THAN 2 LITERS PER 100 KM
ELECTRICITY CHARGING STATIONS
BATTERY AUTONOMY AND POWER
DRIVERLESS VEHICLES
ELECTRIC PLANE AND NEW GENERATION OF EASY-TO-STEER AIRCRAFT AND CIVILIAN DRONES
EMBEDDED SOFTWARE AND SYSTEMS
SATELLITES WITH ELECTRIC PROPULSION
NEXT GENERATION HIGH-SPEED TRAINS
GREEN SHIPS
TECHNICAL AND INTELLIGENT TEXTILES
WOOD INDUSTRIES
RECYCLING AND GREEN MATERIALS
THERMAL RENOVATION OF BUILDINGS
INTELLIGENT ELECTRICITY NETWORKS
WATER QUALITY AND MANAGEMENT OF SCARCITY
GREEN CHEMISTRY AND BIO-FUELS
MEDICAL BIOTECHNOLOGIES
DIGITAL HEALTHCARE
MEDICAL DEVICES AND NEW HEALTHCARE EQUIPMENT
INNOVATIVE PRODUCTS FOR SAFE, HEALTHY AND SUSTAINABLE NUTRITION
BIG DATA
CLOUD COMPUTING
E-LEARNING
TELECOM INDEPENDENCE
NANO-ELECTRONICS
CONNECTED OBJECTS
ENHANCED REALITY
CONTACTLESS SERVICES
SUPERCALCULATORS
ROBOTICS
CYBERSECURITY
FACTORY OF THE FUTURE

Figure 8.4. *The 34 plans for a new industrial France*

Figure 8.5. *Tesla factory – Fremont, USA (source: Tesla Motors)*

OPEO has broadened the debate [OPE 14]. During a presentation made in the annual general meeting of the French Public Investment Bank BPI (Banque Publique d'Investissement – BPIFrance Inno Generation [BPI 16]) last May, which every year brings together the major players in digital transformation, both large firms as well as promising start-ups, David Machenaud and Michaël Valentin defended the topic of operational excellence. A tool for making companies more flexible and economically efficient, the advent of industry 4.0 extensively redefines the very shape of the organization of a company (management in flexible units, communication and horizontal decision-making, the key role of manager/coach), as well as beyond it (transformation of the added value chain with the customer gatecrashing the production unit). For OPEO, the key to the success of factories of the future lies in making the technical, organizational and human aspects consistent.

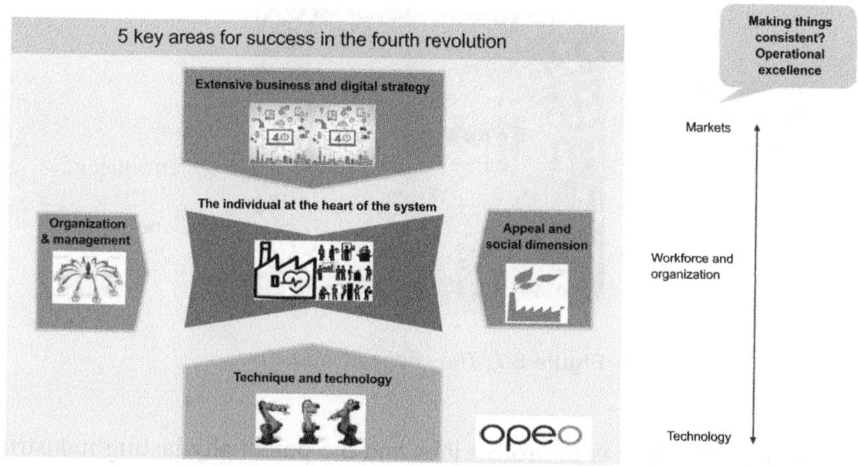

Figure 8.6. *The individual at the heart of operational excellence (source: OPEO)*

In this way, as shown in Figure 8.6, regardless of the managerial or operational level, the individual must make concessions to the four other dimensions [OPE 16]:

understanding production-related techniques and technologies (choosing the most suitable tools, training and increasing the skills of teams, the man–machine relationship);

new modes of organization (progressive abandonment of pyramid structures in favor of autonomous decision-making units);

the attractive dimension of the company (corporate social responsibility favoring the employment of high-quality candidates, with a lower turnover related to optimal working conditions);

meeting demanding and unpredictable markets (real time, "customer centric") through a multi-channel oriented company (e-commerce, new distribution channels, globalized market).

For instance, Patrium, a citizen-led initiative whose aim is to support struggling SMB (Small and Medium-sized Businesses), has taken on a vision in which small industries equipped with digital means are able to take their place at the heart of an uberized economy, directly connecting consumers and producers [PAT 16].

Figure 8.7. *The uberized industry*

The objective here is to protect jobs and the potentially lasting industrial heritage of the territories, which have been put at risk by relocation and factory closures linked to lack of competitiveness associated with the "pre-Internet" age. Thanks to a vision combining digital, collaborative work in SMB networks, and benefiting small-scale, flexible production intended for long-tail markets (niche markets/groups of loyal consumer/actors who are demanding in terms of quality, value and principles, aiming for a global, rather than simply local, scale), Patrium carries the hope of progressive local reindustrialization. Based on 25 years of experience that have enabled the continuation of approximately 50 companies and over 1,000 jobs in fields as varied as fastener manufacturing, textiles, wood and plastics processes, the initiative was consolidated in 2014 with the creation of a Fund for Citizen and Mutual Investment, which currently supports companies throughout the pilot region of Nouvelle-Aquitaine in France. The strong underlying idea is the transmission to future generations of industrial tools that provide jobs. In order to do this, the economic aspect of financing related to the concept of 3SE (Social, Solidary and Solvable Economy), in which there is a return on investments that is as much social and societal as it is (very reasonably) financial. Innovation here lies in combining a vision (potential reindustrialization) with an economic model based on sharing, fundamental to which are tools (digital tools, collaborative tools, production tools providing service) with the aim of transmission (sustainability, jobs).

'Broussaud Textiles'

Les Cars, Limousin, France

Recovered in 2008

300 K€ invested

25 jobs made sustainable

25% growth per year, on average

Box 8.1. *Broussaud Textiles, Patrium test case*

Lastly, a special mention should go to the development of an industry that increasingly favors effective and sustainable waste processing, as well as the repair and maintenance of goods to extend their lifetime. As an example, the start-up Save, which has positioned itself to repair of any electronic object in a market which will have 50 billion connected objects by 2020 and which, in one year, has risen from having an annual turnover of 100K € to a daily turnover of 100K € [SAV 16].

Consequently, innovation can no longer simply be combined with the response to a use-related need or a given systemic problem. It must, fundamentally and in good conscience, address the social and economic issues of our age. Knowing, for example, that 20 kg of raw material serves to produce a chip weighing 0.09 g [ADE 16], 50 billion electronic chips require the use of 10 Gigatons of raw material (in an equivalent "Harmony of the Seas", this represents almost 4,500 liners). This all falls within a recurrence dictated by a renewal frequency for devices for personal use of 5–7 years (taking into account the extremely reasonable assumptions relating to repairs and extended lengths of usage).

This all simply indicates that we have been led to draw on the (*a fortiori* non-renewable) reserves of our planet in a recurrent, and increasing, manner. The case of rare earths is itself the archetype of harmful exploitation. It may soon be followed, or even surpassed, by lithium exploitation. This has been foretold by the Tesla Motors' investment of 5 billion USD in its Gigafactory project in Nevada [TES 16], intended to guarantee the production of lithium-ion batteries for all of the company's cars from 2017, at a pace of production estimated to be 500,000 vehicles per year. Tesla itself prophesied that, "by 2020, the Gigafactory will have reached its maximum capacity and will produce more lithium-ion batteries per year than were produced globally in 2013". This is, once again, concerning, taking into account the related impact [PAR 13].

"Anyone who believes exponential growth can go on forever in a finite world is either a madman or an economist." (Kenneth E. Boulding, quoted in *Jump the Curve* – Jack Uldrich [JAC 08])

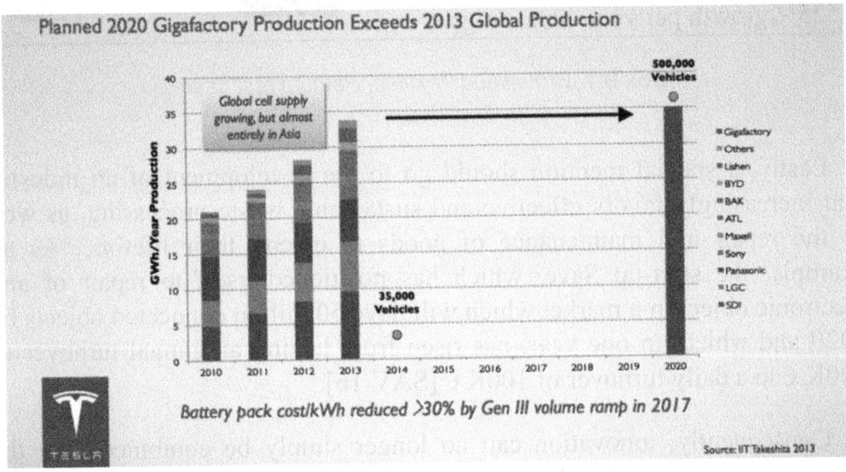

Figure 8.8. *Tesla Gigafactory – planned production of lithium-ion batteries (source: IIT Takeshita, 2013)*

So, what use is an innovation as fabulous as the electric car, intended to cause less pollution (its main selling point), if, at the same time, at the very center of its production process, there are risks of damaging the environment?

Regardless of how many manufacturing processes could have been improved by now (and there are more and more) and attractive green guidelines signed (which should be encouraged), it seems complicated to reverse the current trend: a gradual, massive, and difficult to stop increase in environmental debts.

Subsequently, in order to allow industry X.0 to take root in a truly lasting spatio-temporal trajectory, how can the very manner of conceiving the production of goods and the way of using ensuing services not be evolved? What is the energy and ecological future of an industry that will be confronted with 9.6 billion people by 2050? It is one the most extraordinary challenges that humanity has ever faced …

8.2.2. *Towards a rebirth, but without actually having to die first (or at least without suffering too much)*

There may be a tendency to believe that suffering (or expecting to) is worse than death. Death, to date, remains an unknown quantity, whereas suffering can be felt, communicated and shared. Most of the time, sadly, death is accompanied by suffering. And so, the 2008 crisis resulted in 500,000 cancer deaths [LE 16], while atmospheric pollution put our brains in danger [FOU 14]. As for the planet, exploitation of shale gas would lead to earthquakes [BUS 16] in Oklahoma. Now is the time for us to put an end to pessimism and act.

Changing the rules of the game …

… is to place at the heart of the industrial "field" three high-profile ingredients:

– the circular economy, allowing reuse, reemployment and recycling (not 99%, but …) of 100% of the resources required to manufacture a product, the product itself and the waste generated by the manufacturing process;

– the knowledge economy, allowing the (boundless?) sharing and dissemination of knowledge, know-how and techniques in an increasingly de-compartmentalized manner, with the ability to produce innovations while involving the maximum number of stakeholders;

– a new entrepreneurial state of mind, necessary to renewing the paradigm required by the current situation.

The outcome: Gunter Pauli's blue economy, which suggests not restricting growth (he calls it a limitless growth) and ecology. To what end? it is a matter, quite simply, for the industry to copy nature. This science has a name: biomimicry. One of its most enthusiastic supporters is the researcher Idriss Aberkane, who includes it in a more globally knowledge-based economy [ABE 15]. The aim is basic: nature is a library that contains an astronomic amount of information that is (very) widely underused, if not too often used destructively. We thus most often continue, he says, "to burn down this library instead of simply exploring it and taking inspiration from it, like photocopying a book taken from one of its sections before putting it back in place and thereby preserving biodiversity". These perspectives could create millions of jobs from research to large-scale industrial production, by moving away from the transformation of raw materials and/or the reuse of industrial waste, without any losses!

Figure 8.9. *Idriss Aberkane (left) and Gunter Pauli (source: © François Tancré, Eventpixr & © Rex, Sipa/Keith Mindham)*

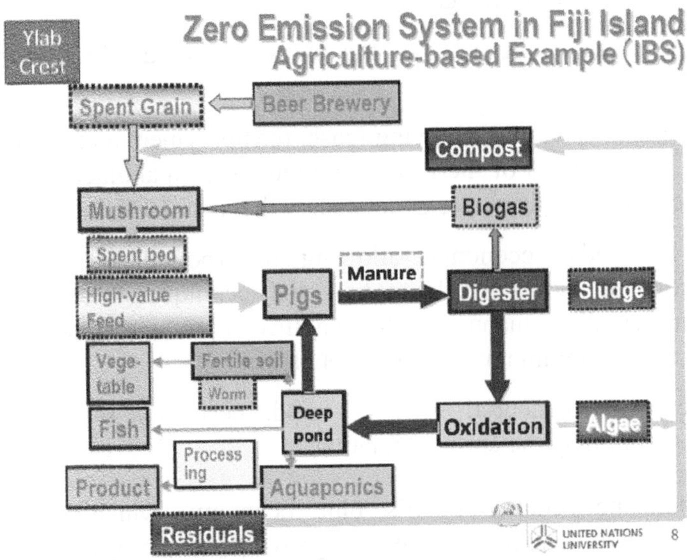

Figure 8.10. *ZERO EMISSION waste processing in a factory ecosystem in Fiji*[3] *(source: Jean Staune)*

3 The waste produced by a brewery is mixed with three kinds of mushrooms that make them edible (for pigs). The excrement produced is injected into a digester that produces biogas, and also green algae (which can be harmful to the environment). However, the algae feed seven species of fish (and a species of crawfish) which digest them in turn.

There are factories operating under these virtuous models [PAU 13]; by taking inspiration from nature, it is, for instance, possible to produce mushrooms from coffee grounds for the textile and agri-food industries. Or even, via a factory in South Africa, to transform stones recovered from quarries into paper! Over 200 companies currently operate according to these principles, a first drop of water in the ocean of industrial production.

So, what precisely are the impacts on innovation processes in large firms and industrial consortiums? It is a matter of taking environmental aspects into consideration from the design stage of the product through the idea of "recoverability". "If a potential waste product (by definition non-recoverable) is present in your product, you should change the composition of your product", warned Jean Staune, citing McDonough and Braungart [MCD 11][4] and Allodi [ALL 13]. We can thus refer to "eco-design".

This opens the door to a transformation of the current economic model, creating a link to so-called functionality economy, within which it will soon no longer be necessary to own an object, but simply to borrow it for determined usage period. Once this time has elapsed, the object is thus recovered by the producer who takes on the responsibility of recycling/repackaging it 100%.

Changing the rules of the game …

… is also, and above all, putting people back at the heart of all economic, societal and innovation-related considerations. People are therefore the operators, the managers (and the coaches), the leaders and the entrepreneurs, as well as those who consume, those who participate in enhancing an ecosystem (familial, societal) and those who have an increasingly global impact (and therefore responsibility) through the means granted to them. This is particularly true at the very heart of companies. It is a question of putting the economy at the service of Man, and not the other way around. It is about innovating by and for Man, whether on an intimate scale (family, friends) or more broadly (at a national, continental or global level). This begins with the self. With the "selves" contextualized in personal spheres, and also in professional and public ones. As Emmanuel Faber, CEO

4 C2C (cradle to cradle).

of Danone, mentioned in an interview with Sister Cécile Renouard [ENT 14], there is only one kind of ethics; company ethics are the same as Ethics with a capital E from the moment that human operators give life to a production system.

For several years, new methods of product management have drawn on the emerging wave of tools that promote an individual and collective consciousness. From liberated companies to holacracy (a system of governance based on collective intelligence), via happiness at work [GUI 16, SHE 15] and the all-new Chief Happiness Officers, all these elements aim at giving individuals a central position in production units and at supporting them in their development. Among these tools, the author enthusiastically refers to the U Theory, which was first shared by ISEP students. Developed by Otto Scharmer (MIT) in 2009 [SCH 16], on the basis of advanced research and numerous field studies, it involves moving through three successive stages to implement new ideas and is deeply subject to a "Meaning":

opening one's mind, being curious, collaborative, an engine for analyzing and processing information (the essential step in any innovation process);

opening one's heart, positioning oneself with kindly dispositions towards oneself and others (colleagues, partners, future users/clients, the environment), integrating within the innovation process fully and tenaciously, an ethical concept of the product/service intended for the "common good";

opening up one's will by experiencing and feeling the "presence", that is, the individual and collective meaning of actions taken, the consequences of what works and what does not that arise from choices made, at different levels, in order to put people (the self and others) at the heart of a harmonious life process for the duration. Then promote taking action.

In a sentence, the author translates this as:

IMPORTANT.– Think Global, Feel Personal, Do Local (but not alone) & Share It.

"We've gone from an Industrial Economy – where we hired hands – to a Knowledge Economy – where we hired heads – to what is now a Global

Human Economy – where we hire hearts."(Dov Seidman, American consultant)

Lastly, changing the rules of the game ...

... is to apply a new set of issues to education and training. Education should be revitalized as for years it has seemed to be on life support, as much in terms of its substance as its form. Using a form of transmission and sharing, particularly by departing from experience, has shown its economic, social and environmental efficiency, all of which is based on the plurality of perceptions of reality that exist today. Through a transfer of values, in the face of threats of rootlessness, supported by strong positions and strong, symbolic gestures, a dream has been revived and a window to hope has been opened.

Education is, in this sense, a duty of utopia, and must be carried by a mission in keeping with the seemingly contradictory needs of our time: pragmatism and idealism, materialism and spirituality, with technical and functional skills being taught as much as knowing how to be, rooted in oneself and in the other, as well as awareness of the spatio-temporal challenges that provides us with the opportunity to endure today and tomorrow.

As for aspects of training (particularly professional), it is a matter of developing it around the aforementioned values, while attempting to be pragmatic and focused on the millions of people who risk finding themselves unemployed, swept aside by the age of digital technology, artificial intelligence [TUA 16] and robotization. Strong guidance and support are required, along with dedicated human and financial resources, acute skills and knowledge in socio-professional environments in different fields (but also between fields!), where each individual who is taken on is valued, when each supporter is given the means to act, where interdisciplinary bridges are created that allow the implementation of systems communicating in terms of expertise and employment. These are to be tested in the relevant territorial contexts, by considering their social, cultural and professional specifications in order to create examples and reproducible case studies.

8.2.3. *Economic alternatives: look left and right before crossing*

Something that has, for some years, been considered utopian, was proposed in a referendum by the Swiss at the beginning of June: an unconditional basic income. The idea is as follows: each citizen benefits from an unconditional allowance regardless of their income, their professional status or their marital status. It amounts to €2250 per adult (50% of an average Swiss salary) and €565 per child. Unsurprisingly, "no" carried the day (with 79%), but for Daniel Häni, it is simply a step along the path and is already a success in itself. "In twenty years, we will have unconditional basic income, maybe even earlier", he told Swiss newspaper Le Temps [ZUN 16]. Knowing that experimental initiatives are rolled out more and more, particularly in northern Europe, it would doubtless be opportune to explore this idea consistently and pragmatically.

Indeed, this could be an extremely impactful initiative, and doubtless a key stage in economic and *a fortiori* industrial development. This measure could contribute to undoing economic and financial pressures that allow people in precarious situations, undergoing career changes, or in training, as well as allowing young people to see connections to work, and life in general, completely differently, even overwhelmingly. Is it not an overwhelming experience that we need today in order to direct the energy of all towards resolving the global challenges that await us, in order to put forward a more humane society, less subject to competition and markets while favoring the emergence of collective intelligence?

However, all of this requires time. We are all aware that, while we work to earn our daily bread, we often have many other things to consider besides saving the planet, reflecting on the meaning of our lives and those of our fellow men and women. Why not, for instance, accompany this "unconditional" income with "incentives" adapted to different contexts? They could thereby promote action by making, for example, a civil or environmental service thanks to this "new time" that would be granted to us, no longer through work, but through activities that define our days, and our lives.

Because yes, in this paradigm, work would gradually disappear in favor of activities that would create initiatives arising from the availability of our skills/knowledge/expertise/individual qualities/information to benefit different companies/associations/structures/factors in the exchange of goods/ services/bitcoin/bartering/local and international currencies recognized as being in the public interest and shared in different economic circles. There is thus no longer any value associated with a job, but *values* associated with the impact created by the activities we carry out. For instance, in different spheres, this is as follows:

– the individual sphere: we carry out an activity for our personal well-being, for our intellectual stimulation, or to take care of our health, in order to feel better in our own lives/for those close to us/in society, to feel happier, consume more, participate, be active and share our happiness;

– the familial sphere: a collective activity in favor of/with our family, our children, parents or grandparents, generating links and intergenerational solidarity;

– the friendly and social sphere: an activity in support of the society in which we live, an investment in our neighborhood or our town, with people who share a "living together" spirit;

– the global sphere: acting for the economy in the sense of *oikonomia* (etymologically, "household management" – meaning, in the present case, the planet), for the environment, openness and intercultural transfer, to implement and promote humanism and ecology.

Each of these activities can be given an assessment of its impact on the basis of the number of hours that have been spent on it and the significance it assumes in a specific context, allowing compensation to be calculated in financial and/or other forms.

We can thus see at which point the industrial project is intertwined, inseparable from the economic (and political) ones, which itself needs to be based on fundamental human values. All of the aspects mentioned are quite clearly far from being exclusive or even exhaustive. Other initiatives, perhaps even more successful, probably exist. But see how difficult, even impossible, it is to decorrelate innovation and industrial progress from a viable economic foundation, without putting Man and its environments at the heart of the system.

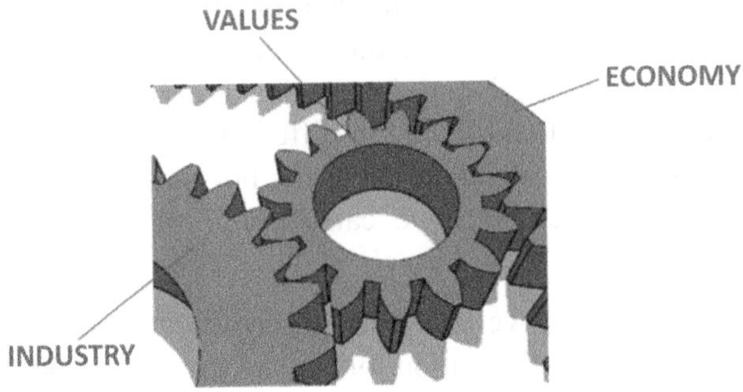

Figure 8.11. *The structure of innovation policies*

In conclusion, we could still ask, while we have technologies, skills, knowledge and expertise, methods and tools, while we make balance sheets, and while, *a priori*, we learn from our mistakes, and even an embryo of global awareness seems to emerge, we remain either trapped, helpless, disinterested, or worried by the urgency of the short term, or unconcerned, even strangers (alien) to these problems. There are of course blocking factors and aggravating factors that are deeply rooted in the collective unconscious and the idea is not to list them or explain them, but to quite simply refute the question! Because the solution is certainly not there! And so, replace this question with another: are we able to create and support a new generation of explorers, entrepreneurs, inventors, innovators and influencers that can profoundly and lastingly integrate these challenges, experiment on a large scale and thus validate the proof of concepts so as to drag rootless populations in their wake?

8.3. Corollaries: impacts on innovation

8.3.1. *Cultures*

The idea here is to promote the emergence of not one, but a number of different innovation cultures, adapted and optimized according to each context (company – start-up, SMB, multinationals, industry, economic model, political system) with the necessary requirement to be linked to a fixed anchor point: the individual, above all and at all times. All innovation must thus justify its reason for being through its capacity to be built with, around and for Man, today and beyond. Given current issues, initiatives allowing the "reimbursement" (a combination of "quick-wins" and long-term strategic actions) of economic and ecological debts that we have set up must be encouraged, instead of simply avoiding aggravating them "too much".

8.3.2. *Visions and strategies*

These cultures must be embodied and carried out by decision makers aware of the stakes and of their responsibilities, who are able to overturn the established divisions and replace competitiveness and excess profit logics (where human challenges are addressed only partially) with creativity, co-production, sharing (of ideas, knowledge and values generated) via multidisciplinary and transgenerational implications. Innovation is represented here by a change in the state of mind and by the completion of experiments in a process of Feeling → Awareness → Verbalization → Decision → Action reaching (a degree of) maturity, and inclined to question itself, evolve and improve in an agile and continuous manner.

8.3.3. *Processes*

Relying on mechanisms that largely exist in different spheres (some of which have been mentioned previously), there is a question of adapting to the restrictions of one's own organizations (individual, family, business, society, world). Innovation resides more and more in the capacity to sort, link and analyze the information available to us in order to select it and combine it in the most optimal manner possible with the aim of, finally, taking the risk of transposing it to its own environments, and finding arguments to convince our partners of the usefulness of doing it!

8.3.4. *Tools*

Lastly, it is about promoting and using tools in order to demonstrate the feasibility of the innovations undertaken. Even then, these tools already exist, whether they are managerial or technological, for the most part. It is our duty to give ourselves the means for our ambitions by detecting them, mastering them and assigning them a usage in accordance with the stated culture, vision/strategy and process.

8.4. Conclusion

After having analyzed, in the first section, the (mostly adverse) effects of industrial policies on the three key areas of human evolution – the economy, society and the environment – we presented three axes of reflection (technological, social and societal values, economic alternatives) that must contribute to a change in the industrial approach, in terms of provoking the jump-start necessary for envisaging the future. Addressing the issue of the sustainability of industry cannot succeed in the event that one of these axes is left aside. It therefore seems essential to begin work immediately on this tripartite evolution.

This is all thanks to innovation, the engine behind all developmental processes. Nonetheless, the specific case of industry highlights, given the urgency of the situation, the need to act to different levels in an open and coordinated way, with a non-negotiable prerequisite: to change the very foundations of the intrinsic cultures and paradigms of innovation by putting Man *back* at the center of the stakes. In two words: deeply rethinking the modes of production, as well as the usefulness of the proposed goods and services.

Already having an impressive array of tools, means and mechanisms from collective human and natural intelligence, the main challenge of innovation is in the application of these potential solutions and their large-scale acceptance by decision makers (largely majority shareholders) from the "postmodern world".

Bringing about new entrepreneurial, industrial, and economic practices in an extremely short period of time depends on social, economic, and environmental contours reshaping the world of tomorrow to an extent never before seen in history.

8.5. Bibliography

[ABE 15] ABERKANE I., Consulted 05.09.2016, available at: http://www.dailymotion.com/video/x2joni8, 2015.

[ADE 16] ADEME, Consulted 05.09.2016, available at: http://www.ademe.fr/, 2016.

[ALL 13] ALLODI E., Consulted 05.09.2016, available at: https://www.youtube.com/watch?v=mvznik6tG4E, 2013.

[BAN 16] BANQUE-MONDIALE, Consulted 05.09.2016, available at: http://donnees.banquemondiale.org/indicateur/NV.IND.TOTL.ZS?end=2014&start=1995&view=chart, 2016.

[BAU 12] BAUMARD P., *Le vide stratégique,* Editions CNRS, Paris, 2012.

[BAU 14] BAUMARD P., Consulted 05.09.2016. on Xerfi Canal: available at: https/www.youtube.com/watch?v=9laVBd5iDgU, 2014.

[BPI 16] BPI-FRANCE, Consulted 05.09.2016, available at: http://big.bpifrance.fr/, 2016.

[BUS 16] BUSSARD S., Consulted 05.09.2016, available at: on LE-MONDE.fr: http://www.lemonde.fr/planete/article/2016/01/15/dans-l-oklahoma-le-gaz-de-schiste-provoque-des-seismes-a-repetition_4847904_3244.html, 2016.

[CEB 16] CEBIT, Consulted 05.09.2016, available at: on www.cebit.de/home, 2016.

[DEN 14] DENIS J.-P., Consulted 05.09.2016, available at: on Xerfi Canal: https://www.youtube.com/watch?v=A339gxxkccU, 2014.

[ENT 14] ENTRETIENS-DE-VALPRE, Consulted 05.09.2016, available at: on https://www.youtube.com/watch?v=hSV4yvMdIZ0&feature=youtu.be&t=21m33s, 2014.

[FOU 14] FOUCART S., Consulted 05.09.2016, available at: on http://www.lemonde.fr/medecine/article/2014/12/10/la-pollution-met-en-danger-le-cerveau_4538177_1650718.html, 2014.

[GAD 14] GADREY J., Consulted 05.09.2016, available at: on http://alternatives-economiques.fr/blogs/gadrey/2014/03/10/210-milliards-d%E2%80%99euros-de-dividendes-oui-mais%E2%80%A6/, 2014.

[GOU 14] GOUVERNEMENT, Consulted 05.09.2016, available at: http://www.economie.gouv.fr/files/files/PDF/nouvelle-france-industrielle-sept-2014.pdf, 2014.

[GOU 16] GOUVERNEMENT, Consulted 05.09.2016, available at: http://www.gouvernement.fr/partage/3813-l-industrie-en-france, 2016.

[GUI 16] GUILLAUD H., Consulted 05.09.2016, on LE-MONDE.fr, 2016.

[INS 16] INSEE, Consulted 05.09.2016, available at: http://www.insee.fr/fr/themes/info-rapide.asp?id=40, 2016.

[IPC 14] IPCC, "Climate Change", UNEP, 2014.

[KAI 14] KAIMAN J., Consulted 05.09.2016, available at: on theguardian: https://www.theguardian.com/sustainable-business/rare-earth-mining-china-social-environmental-costs, 2014.

[LAL 14] LALOUX F., *Reinventing Organizations*, Editions Nelson Parker, 2014.

[LAT 16] LATRIBUNE.fr., Consulted 05.09.2016, available at: on La Tribune: http://www.latribune.fr/economie/international/la-4e-revolution-industrielle-menace-plus-de-5-millions-d-emplois-543777.html, 2016.

[LE 16] LE-MONDE.fr., Consulted 05.09.2016, available at: on http://www.lemonde.fr/sante/article/2016/05/26/plus-de-500-000-morts-par-cancer-dans-le-monde-associees-a-la-crise-de-2008_4926524_1651302.html, 2016.

[MAT 15] MATTEI J.-F., *L'homme dévasté*, Editions Grasset, 2015.

[MCD 11] MCDONOUGH W., BRAUNGART M., *Cradle to Cradle*, Editions Manifestô, 2011.

[MOR 11] MORIN E., *La Voie*, Editions Fayard, 2011.

[OPE 14] OPEO, Consulted 05.09.2016, available at: on http://www.opeo-conseil.fr/, 2014.

[OPE 16] OPEO, Consulted 05.09.2016, available at: on http://www.challenges.fr/tribunes/20160719.CHA1969/quelle-sera-la-france-industrielle-de-2025.html, 2016.

[OXF 15] OXFAM, Insatiable richesse, toujours plus pour ceux qui ont déjà tout, available at: http://www.oxfamfrance.org/sites/default/files/file_attachments/rapport_oxfam_insatiable_richesse_0.pdf, 2015.

[PAR 13] PARASOTE V., Consulted 05.09.2016, available at: on http://twi-terre.net/index.php/infos-sourcees/55-batterie-lithium-ion-un-poids-plume-aux-lourds-impacts, 2013.

[PAT 16] PATRIUM, Consulted 05.09.2016, available at: on http://www.patrium.fr/, 2016.

[PAU 13] PAULI G., Consulted 05.09.2016, available at: on TEDx: http://www.dailymotion.com/video/x2joni8, 2013.

[PAU 15] PAULI G., Consulted 05.09.2016, available at: on http://www.theblueeconomy.org/, 2015.

[RIE 11] RIES E., *Lean Startup*, Editions Pearson, 2011.

[SAL 15] SALMON K., Consulted 05.09.2016, available at: on http://www.kurtsalmon.com/uploads/2015_POV_%20U%26I_Indutrie%204.0_WEB_VF.pdf, 2015.

[SAV 16] SAVE, Consulted 05.09.2016, available at: on https://www.save.co/fr/fr, 2016.

[SCH 16] SCHARMER O., *Theory U – Leading from the Future as it Emerges*, 2nd ed., Berret-Koeler Publishers, 2016.

[SHE 15] SHERIDAN R., *Joy, Inc.*, Editions Penguin, 2015.

[STA 15] STAUNE J., *Les clés du Futur,* Editions Plon, 2015.

[STA 16] STAUNE J., Consulted 05.09.2016, available at: on https://www.youtube.com/watch?v=NZKqPoQiaDE&feature=youtu.be&t=14m2s, 2016.

[SUP 15] SUPIOT A., *La gouvernance par les nombres*, Editions Fayard, 2015.

[TES 16] TESLA, Consulted 09 06, 2016, available at: sur https://www.tesla.com/fr_FR/gigafactory, 2016.

[TUA 16] TUAL M., Consulted 05.09.2016. available at: on LE-MONDE.fr: http://www.lemonde.fr/pixels/article/2016/06/17/intelligence-artificielle-google-lance-un-groupe-de-recherche-europeen-sur-l-apprentissage_4952666_4408996.html, 2016.

[ZUN 16] ZUND C., Consulted 05.09.2016, available at: on LE-TEMPS: https://www.letemps.ch/suisse/2016/03/18/entrepreneur-lance-initiative-un-revenu-base-inconditionnel, 2016.

List of Authors

Audrey DEPEIGE
Telecom Business School
Évry
France

Theodor FELEZEU
ISEP Formation Continue
France

Marc JAILLOT
University of French Polynesia
Faaa
French Polynesia

Yann DE KERMADEC
Groupe Julhiet
France

Gérard A. KOKOU DOKOU
Research on Management and Economics
University of Littoral Côte d'Opale
Dunkerque
France

Michel MARCHESNAY
University of Montpellier
France

Maggy PERRIER
Essilor International
Charenton-le-Pont
France

Pierre SAULAIS
Institut Mines-Telecom
Paris
France

Corinne TANGUY
University of Bourgogne
Franche-Comté
INRA AgroSup Dijon
Research Network on Innovation
France

Dimitri UZUNIDIS
Research Unit on Industry and
Innovation
University of Littoral Côte
d'Opale
Research Network on Innovation
France

Index

A, B, C

D, E, F

H, I, K

M, N, O

P, R, S

T, V

Other titles from

iSTE

in

Innovation, Entrepreneurship and Management

2017

CHALENÇON Ludivine
Location Strategies and Value Creation of International Mergers and Acquisitions

LEBERT Didier, EL YOUNSI Hafida
International Specialization Dynamics (Smart Innovation Set – Volume 9)

MEDINA Mercedes, HERRERO Mónica, URGELLÉS Alicia
Current and Emerging Issues in the Audiovisual Industry (Diverse and Global Perspectives on Value Creation Set – Volume 1)

SAVIGNAC Emmanuelle
The Gamification of Work: The Use of Games in the Workplace

2016

BARBAROUX Pierre, ATTOUR Amel, SCHENK Eric
Knowledge Management and Innovation (Smart Innovation Set – Volume 6)

BEN BOUHENI Faten, AMMI Chantal, LEVY Aldo
Banking Governance, Performance And Risk-Taking: Conventional Banks Vs Islamic Banks

BOUTILLIER Sophie, CARRÉ Denis, LEVRATTO Nadine
Entrepreneurial Ecosystems (Smart Innovation Set – Volume 2)

BOUTILLIER Sophie, UZUNIDIS Dimitri
The Entrepreneur (Smart Innovation Set – Volume 8)

BOUVARD Patricia, SUZANNE Hervé
Collective Intelligence Development in Business

GALLAUD Delphine, LAPERCHE Blandine
Circular Economy, Industrial Ecology and Short Supply Chains (Smart Innovation Set – Volume 4)

GUERRIER Claudine
Security and Privacy in the Digital Era (Innovation and Technology Set – Volume 1)

MEGHOUAR HICHAM
Corporate Takeover Targets

MONINO Jean-Louis, SEDKAOUI Soraya
Big Data, Open Data and Data Development (Smart Innovation Set – Volume 3)

MOREL Laure, LE ROUX Serge
Fab Labs: Innovative User (Smart Innovation Set – Volume 5)

PICARD Fabienne, TANGUY Corinne
Innovations and Techno-ecological Transition (Smart Innovation Set – Volume 7)

2015

CASADELLA Vanessa, LIU Zeting, DIMITRI Uzunidis
Innovation Capabilities and Economic Development in Open Economies (Smart Innovation Set – Volume 1)

CORSI Patrick, MORIN Dominique
Sequencing Apple's DNA

CORSI Patrick, NEAU Erwan
Innovation Capability Maturity Model

FAIVRE-TAVIGNOT Bénédicte
Social Business and Base of the Pyramid

GODÉ Cécile
Team Coordination in Extreme Environments

MAILLARD Pierre
Competitive Quality and Innovation

MASSOTTE Pierre, CORSI Patrick
Operationalizing Sustainability

MASSOTTE Pierre, CORSI Patrick
Sustainability Calling

2014

DUBÉ Jean, LEGROS Diègo
Spatial Econometrics Using Microdata

LESCA Humbert, LESCA Nicolas
Strategic Decisions and Weak Signals

2013

HABART-CORLOSQUET Marine, JANSSEN Jacques, MANCA Raimondo
VaR Methodology for Non-Gaussian Finance

2012

DAL PONT Jean-Pierre
Process Engineering and Industrial Management

MAILLARD Pierre
Competitive Quality Strategies

POMEROL Jean-Charles
Decision-Making and Action

SZYLAR Christian
UCITS Handbook

2011

LESCA Nicolas
Environmental Scanning and Sustainable Development

LESCA Nicolas, LESCA Humbert
Weak Signals for Strategic Intelligence: Anticipation Tool for Managers

MERCIER-LAURENT Eunika
Innovation Ecosystems

2010

SZYLAR Christian
Risk Management under UCITS III/IV

2009

COHEN Corine
Business Intelligence

ZANINETTI Jean-Marc
Sustainable Development in the USA

2008

CORSI Patrick, DULIEU Mike
The Marketing of Technology Intensive Products and Services

DZEVER Sam, JAUSSAUD Jacques, ANDREOSSO Bernadette
Evolving Corporate Structures and Cultures in Asia / Impact of Globalization

2007

AMMI Chantal
Global Consumer Behavior

2006

BOUGHZALA Imed, ERMINE Jean-Louis
Trends in Enterprise Knowledge Management

CORSI Patrick *et al.*
Innovation Engineering: the Power of Intangible Networks

Printed and bound by CPI Group (UK) Ltd, Croydon, CR0 4YY
16/09/2026
14963072-0001